Three Things to Help Heal the Planet

About the Author

Ana Santi is a journalist, writer and editor. She has written for publications such as *The Times* and the *Guardian* and been a regular commentator for the BBC, *Sky News* and *Woman's Hour*. She is the co-editor of *Comfort Zones*, an anthology featuring original works by 28 women writers, and volunteers as a writing mentor with the Ministry of Stories, supporting children's literacy. She lives in London.

Three Things to Help Heal the Planet

21 Simple Ideas from Environmental Trailblazers So You Can Make a Difference Today

Ana Santi

Published in 2022 by Welbeck Balance
An imprint of Welbeck Non-Fiction Ltd
Part of Welbeck Publishing Group
Based in London and Sydney.
www.welbeckpublishing.com

A CIP catalogue record for this book is available from the British Library

ISBN
Paperback – 978-1-80129-075-3

Typeset by Lapiz Digital Services
Printed in Great Britain by CPI Group (UK) Ltd, Croydon CRO 4YY

10 9 8 7 6 5 4 3 2 1

Note/Disclaimer

Welbeck Balance encourages diversity and different viewpoints. However, all views, thoughts, and opinions expressed in this book are the author's own and are not necessarily representative of Welbeck Publishing Group as an organization. All material in this book is set out in good faith for general guidance; Welbeck Publishing Group makes no representations or warranties of any kind, express or implied, with respect to the accuracy, completeness, suitability or currency of the contents of this book, and specifically disclaims, to the extent permitted by law, any implied warranties of merchantability or fitness for a particular purpose and any injury, illness, damage, death, liability or loss incurred, directly or indirectly from the use or application of any of the information contained in this book. This book is not intended to replace expert medical or psychiatric advice. It is intended for informational purposes only and for your own personal use and guidance. It is not intended to diagnose, treat or act as a substitute for professional medical advice. The author and the publisher are not medical practitioners nor counsellors, and professional advice should be sought before embarking on any health-related programme.

www.welbeckpublishing.com

For Stella

Introduction: First Things First xiii

Meet the Experts xxi

LIVE 1

1. JB MacKinnon
   The Day the World Stops Shopping 7
2. Khandiz Joni
   The Bathroom Epiphany 17
3. Claire Ratinon
   Grow Food, Dismantle the System 27

Three Things to Do 32

EAT 35

4. Tristram Stuart
   Throw a Better Party 41
5. Melissa Hemsley
   In It Together 45
6. Jonathan Safran Foer
   We Are the Weather: Saving the Planet Begins at Breakfast 49

Three Things to Do 54

WEAR 57
7. Eshita Kabra-Davies
"If Not Me, Who? If Not Now, When?" 65
8. Tamsin Blanchard
Lessons from My Mother 69
9. Christopher Raeburn
Make Good Use of Bad Rubbish 73
Three Things to Do 78

TRAVEL 81
10. Rosamund Kissi-Debrah
A Breath of Fresh Air 85
11. Juliet Kinsman
On How to Travel 89
12. Jona Christians
Motoring toward New Ideas 95
Three Things to Do 100

SPEND 103
13. Alasdair Roxburgh
Money Matters 109
14. Tessa Wernink
Explore the Alternative 117
15. Catherine Chong
Investments with Impact 125
Three Things to Do 131

BREATHE 135

16. Dr Imogen Napper
Turning Ripples into Waves 141

17. Yvonne Aki-Sawyerr, OBE
Breathe the Change 147

18. Dr Emily Shuckburgh, OBE
Helping the Planet from Home 153

Three Things to Do 159

TRANSFORM 163

19. Dr Gail Bradbrook
From Grief Comes Courage 169

20. Lucy von Sturmer
Challenge Accepted 173

21. Jennifer Martel
The Bigger Picture 181

Three Things to Do 186

Epilogue 189

The 21 Solutions 193

Acknowledgements 197

References 199

Resources 215

Introduction: First Things First

I know the air I breathe is polluted, that the fish I eat live on a diet of plastic, that a coat I don't need shouldn't cost £30. But for a long time, I didn't know what to do about it. Not for want of information – I was drowning in news reports, opinion pieces, podcasts and Instagram posts – but for want of direction: endless debate and contradictory solutions, if indeed solutions were offered at all. What if I could get the scientists, activists and entrepreneurs – global experts in their fields – to categorically tell me three things that, through collective action, could slow down the rate of greenhouse gas emissions from the food industry? Three things that, if tackled as a community, could reduce air pollution. Three things to make our wardrobes more sustainable.

When I began my journalism career, I was taught the rule of three. Three makes a trend. Always have at least three corroborative sources before writing a story. It works in classic storytelling, too: "The Three Little Pigs", "Goldilocks and the Three Bears", "The Three Billy Goats Gruff". There is something simple and satisfying about three. The repetition of three is manageable and memorable.

As a journalist, I hold people to account, distil information and present it without fuss. *Three Things to Help Heal the Planet* is the story of a to-do list; a collection of 21 essays from experts around the world on how to collectively tackle the climate and ecological crises. Each personal story concludes with a call to action.

Part of the argument against simplification is that the climate crisis is too complicated; individual action alone isn't enough to prevent a catastrophe. I agree. We need to do a lot more than cycle to work and recycle our newspapers – and we need to do it now. So in that respect, I admit *Three Things* is flawed. But so am I, so are you. The actions we're each prepared to take will vary significantly depending on our circumstance and character, yet most of us are united in wanting to do something. In their book *The Future We Choose, Surviving the Climate Crisis,*

Christiana Figueres and Tom Rivett-Carnac propose two extreme groups of people: climate deniers and those who accept the science but are losing confidence that anything can be done about it: "A larger group of people, between these two extremes, understand the science and acknowledge the evidence but take no action because they don't know what to do."[1]

"If you dare to care, it's a head-fuck", Lily Cole writes in *Who Cares Wins: Reasons for Optimism in our Changing World*. "One day, biofuels are heralded as our saviour: another day they are driving deforestation in Indonesia. How much do we need individual change; how much system change? Where does one end and the other begin? How powerful are we? How responsible are we? Sometimes it feels like you can't even breathe without stealing oxygen."[2]

Three Things to Help Heal the Planet is a first step to owning the problem. The more actions we take – however small at the beginning – the more likely we are to engage in the bigger, necessary efforts. As the scientist Tom Crowther said in a TED talk, dismissing tangible actions is "ultimately an excuse to do nothing. 'Oh, if we can't achieve 100 per cent let's not bother.'"[3] *Three Things*, by definition, cannot be an exhaustive list. Instead, it's an argument against endless debate in favour of decisive action: stop worrying, start doing.

And let's not forget the power we hold as individuals when we come together. Until the spring of 2020, references to the collective human spirit typically focused on World War II. British prime minister Boris Johnson, in his self-appointed role of Winston Churchill, revelled in wartime rhetoric when addressing the nation at the start of the Covid-19 pandemic, before people in power broke the very rules they'd enforced. In those early lockdown days, the coronavirus gave us a modern history of collective endeavour. One man baked bread for his entire street. Within hours of a request, half a million people signed up to volunteer for the NHS.[4] We all stepped off the pavement to allow others to pass at a safe distance.

So how do we come together over the climate crisis? We've already started. We've reduced our daily meat consumption in the UK by 17 per cent, compared to ten years ago.[5] Given that the global food system is responsible for 30 per cent of all greenhouse gas emissions, eating less meat collectively drives systemic change.

Three Things examines our relationship with the everyday: food, clothes and technology; the air we breathe, the institutions we bank with, the businesses we buy from. All things that have a direct impact on our planet. The fashion industry contributes to around 10 per cent of global carbon emissions. The world's biggest banks have provided $3.8tn of financing for fossil fuel companies since 2015. Our oceans will contain more plastic than fish by 2050 if we do nothing. Figueres and Rivett-Carnac sum it up best: "The goal of halving global emissions by 2030 represents the absolute minimum we must achieve if we are to have at least a 50% chance of safeguarding humanity from the worst impacts. We are in a critical decade."[6]

My biggest fear in turning a personal to-do list into this book was how to tackle the depressing nature of these stats. I needn't have worried. Every essay in this collection moved me to action. The first expert I spoke to was Tristram Stuart. His response? "I intend to live a good life while fighting the good fight." With Melissa Hemsley, I kept hoping she'd invite me round for dinner. Tamsin Blanchard unwittingly took me back to my childhood. Christopher Raeburn made me laugh.

Others made me cry – with hope and ambition. Rosamund Kissi-Debrah, Dr Gail Bradbrook, Mayor Yvonne Aki-Sawyerr, Jennifer Martel, Catherine Chong – theirs are stories of courage, of turning dissatisfaction into action. Doctors Emily Shuckburgh and Imogen Napper took me to places of wonder, to the polar regions and the Ganges. I changed my mind with narratives of innovation from JB MacKinnon, Khandiz Joni, Lucy von Sturmer, Tessa Wernink and Jona Christians.

Then there were the stories of contradiction. The journalist Juliet Kinsman who, after decades of a peripatetic lifestyle, became a leading sustainable travel expert. The Chair of Friends of the Earth Europe and advocate for ethical banking, Alasdair Roxburgh, who chose his first bank based on its offer of a free railcard. Eshita Kabra-Davies, a self-styled consumerist turned fashion rental entrepreneur. The organic food grower Claire Ratinon who, until ten years ago, had never seen food being grown. The author Jonathan Safran Foer who, in making the case that factory-farmed meat should not be eaten, consumed some bad burgers. You don't have to be a saint to save the world.

This book is for anyone whose everyday lives are rich in choice. Anyone in a position to choose what they eat and wear, how they travel and live, where they bank and shop and who they vote for, has power. This book is for everyone who wants to use that power to make a difference.

Each chapter in *Three Things* has three essays and concludes with a trio of actions. A total of 21 solutions. Some couldn't be simpler: buy fewer things. Others require more effort: would you quit your job if your employer's values didn't align with your own? *Three Things to Help Heal the Planet* begins with the simpler, more tangible actions – although personal circumstance will ultimately dictate the hierarchy. As it progresses, the book's calls to arms become more nuanced; they call for a change in mindset.

Perhaps you'll do three things from one chapter. Perhaps you'll select three from different chapters. Perhaps you'll do them all. You choose – everything is connected.

Perhaps you'll add to the list. I offer the 21 solutions as a menu; a starting point for anyone who, like me, is looking for tangible ways to make a difference. I don't doubt there will be additions in the coming weeks, months and years. I imagine there'll be revisions, too. But when in doubt, I'm reminded of Lily Cole's words: "I'm not interested in ideological fights, or claims to right and wrong. I am interested in building bridges; in engaged, proactive optimism."[7]

I'm interested in action.

We each get our inspiration and motivation from all sorts of places. I love tennis. The former player Billie Jean King, who fought for equality on and off the court, once said – and I paraphrase – that you should leave the sport in a better place than you found it. I've been lucky to live in a world full of opportunity, on a beautiful planet. I owe it to my daughter and future generations to give them the same chance. The people most impacted by the climate crisis are often the least responsible for it. Those of us who can do something absolutely should.

Meet the Experts

From activists, authors and entrepreneurs to scientists, designers and politicians, meet the 21 experts across the world and read their stories for collective action against the climate emergency.

YVONNE AKI-SAWYERR, OBE

Chapter 17, Breathe the Change

Yvonne Aki-Sawyerr is the mayor of Freetown, the capital city of Sierra Leone. In 2015, she was recognized for her work during Sierra Leone's Ebola crisis with an Ebola Gold Medal by the then President of Sierra Leone, Ernest Bai Koroma. A year later, she was made an Officer of the Order of the British Empire (OBE). As mayor, Yvonne launched the Transform Freetown initiative, which includes the campaign to plant a million trees in two years.

TAMSIN BLANCHARD

Chapter 8, Lessons from My Mother

Tamsin Blanchard is a fashion journalist, editor and author, whose books include *Green is the New Black: How to Save the World in Style*. She is part of the global not-for-profit organization Fashion Revolution and curator of Fashion Open Studio, a series of events to celebrate the people and processes behind the making of our clothes.

DR GAIL BRADBROOK

Chapter 19, From Grief Comes Courage

Gail Bradbrook is a co-founder of the social movement Extinction Rebellion (XR), which has rapidly spread internationally since its launch in 2018. She has a PhD in molecular biophysics.

CATHERINE CHONG

Chapter 15, Investments with Impact

Catherine Chong is a scientist and ESG (ecology, social, and governance) advisor via her company LinkESG. In 2020, she co-founded Farms To Feed Us, a UK-wide database of farms, and has been lobbying for a mandatory method-of-production food labelling in the UK with CLEAR. She is now developing a circular platform to support large-scale farming and forestry transitioning to agroecological and regenerative farming.

JONA CHRISTIANS

Chapter 12, Motoring toward New Ideas

Jona Christians is the co-founder and CEO of Sono Motors, a solar and electric car company based in Germany. In 2016, he co-founded the company with Laurin Hahn and Navina Pernsteiner. Together with Laurin, Jona manually laid the foundation for the company's first vehicle, the Sion. His main areas of expertise are the company's product and innovation strategy.

MELISSA HEMSLEY

Chapter 5, In It Together

Melissa Hemsley is a self-taught chef, author and sustainability champion. She is an ambassador for Fairtrade UK and has written four cookbooks, including *Eat Green,* which features the dishes mentioned in her essay.

ESHITA KABRA-DAVIES

Chapter 7, "If Not Me, Who? If Not Now, When?"

Eshita Kabra-Davies is the founder and CEO of award-winning app By Rotation, the UK's first peer-to-peer fashion rental platform. In September 2020, By Rotation joined Sharing Economy UK, the trade body representing the UK's sharing economy businesses, where Eshita is a board member.

KHANDIZ JONI

Chapter 2, The Bathroom Epiphany

Khandiz Joni is a former professional hair and make-up artist turned Creative Sustainableist. She is a founding member of Conscious Beauty Union, an online resource of conscious personal care brands. Khandiz uses creativity, humour and empathy to inspire behavioural change in individuals and business leaders. Her favourite colour is yellow (except if she has to wear it, in which case it's a terrible colour).

JULIET KINSMAN

Chapter 11, On How to Travel

Juliet Kinsman is a journalist, editor, consultant and author of *The Green Edit: Travel*. She is the founder of Bouteco, a collection of boutique and eco hotels, and is Sustainability Editor for *Condé Nast Traveller*.

ROSAMUND KISSI-DEBRAH

Chapter 10, A Breath of Fresh Air

Rosamund Kissi-Debrah is the founder of the Ella Roberta Family Foundation. In 2020, a coroner made legal history by ruling that air pollution was a contributing factor to her daughter's death. At the time of writing, Ella is the only person in the world to have air pollution listed on her death certificate. Rosamund's campaigning has gained worldwide recognition and she is the World Health Organization Breathe Life Ambassador.

JB MACKINNON

Chapter 1, The Day the World Stops Shopping

JB MacKinnon is the author or co-author of five books of nonfiction, including his latest, *The Day the World Stops Shopping*, a thought experiment that imagines what would happen to our economies, our products, our planet, ourselves if we committed to consuming far fewer of the Earth's resources. An award-winning journalist, his work has appeared in *The New Yorker*, *National Geographic* and *The Atlantic*.

JENNIFER MARTEL

Chapter 21, The Bigger Picture

Jennifer Martel is a board member of the Indigenous Peoples Movement, which unites Indigenous peoples from across the world to bring awareness of the issues affecting their communities. Jennifer is the Visitor Coordinator of the Sitting Bull Visitor Center at Sitting Bull College in North Dakota, and has been planning, organizing and implementing arts and food classes for seven years in Standing Rock to help revitalize communities.

DR IMOGEN NAPPER

Chapter 16, Turning Ripples into Waves

Imogen Napper is a postdoctoral researcher at the University of Plymouth in the UK. Her research investigates the sources of plastic pollution into the environment.

CHRISTOPHER RAEBURN

Chapter 9, Make Good Use of Bad Rubbish

Christopher Raeburn is the founder of fashion brand RÆBURN, whose ethos has pioneered the reworking of surplus fabrics and garments to create distinctive and functional clothes. He is also the Global Creative Director of Timberland.

CLAIRE RATINON

Chapter 3, Grow Food, Dismantle the System

Claire Ratinon is an organic food grower and writer based in East Sussex. Her work ranges from growing produce for Ottolenghi restaurant, Rovi, to delivering growing workshops in primary schools. Claire's first book is *How To Grow Your Dinner Without Leaving the House*. Her second is *Unearthed: On race and roots, and how the soil taught me I belong.*

ALASDAIR ROXBURGH

Chapter 13, Money Matters

Alasdair Roxburgh is Chair of Friends of the Earth, Europe. He has spent much of his career working for the third sector, including Save the Children and Christian Aid, where he has campaigned on issues including tax justice, child refugee rights and the climate emergency.

JONATHAN SAFRAN FOER

Chapter 6, We Are the Weather: Saving the Planet Begins at Breakfast

Jonathan Safran Foer is the author of the widely acclaimed *Eating Animals*, an investigation into the ethics of animal agriculture. His essay is an extract from his latest work of nonfiction, *We Are The Weather: Saving the Planet Begins at Breakfast.*

DR EMILY SHUCKBURGH, OBE

Chapter 18, Helping the Planet from Home

Emily Shuckburgh is a world-leading climate scientist and the Director of Cambridge Zero, a climate change initiative from Cambridge University to create solutions for a zero-carbon future. Emily is also Professor in Environmental Data Science in the Department of Computer Science and Technology. In her previous role at the British Antarctic Survey, she led a national research programme on polar climate change.

TRISTRAM STUART

Chapter 4, Throw a Better Party

Tristram Stuart is an award-winning author, campaigner and expert on the environmental and social impacts of food. He is the founder of campaign group Feedback, and Toast Ale, a beer made using fresh surplus bread.

TESSA WERNINK

Chapter 14, Explore the Alternative

Tessa Wernink is a social entrepreneur, business coach and campaigner. She supports purpose-driven entrepreneurs who are questioning current systems and developing more ethical and sustainable alternatives. She creates stories through the podcast series *What If We Get It Right?* and is co-founder of The Undercover Activist, a platform that offers courses on positive workplace activism. She is co-founder of the social enterprise Fairphone.

LUCY VON STURMER

Chapter 20, Challenge Accepted

Lucy von Sturmer is the founder of award-winning impact consultancy The Humblebrag, which supports business leaders and brands to take a stand on social, cultural and environmental issues. Lucy is also the initiator of Creatives for Climate, a global network of more than 2,000 creatives, activists, policy-makers and passionate humans working collaboratively to drive action and awareness on the climate emergency.

- People in rich countries consume up to ten times more natural resources than those in the poorest countries. On average, a North American consumes around 90kg of resources each day; a European 45kg per day; while in Africa people consume only around 10kg per day.[1]
- The beauty industry produces 120 billion units of packaging per year. Water is normally the leading ingredient in beauty products: the average moisturiser contains 60 per cent water, while liquid shampoos and conditioners comprise about 80 per cent water. Meanwhile, one in three people globally lack access to clean drinking water. The beauty industry is a major contributor to water pollution.[2]
- In the US, for every $1 spent on food by the consumer, only 11 cents goes to agricultural production. The same is true in the UK, where agriculture accounts for only 10 per cent of the value of the food system.[3]

When I was little, my dad kept a cardboard box informally known as the Doll's Hospital. In it, you'd find plastic arms and legs, the occasional head, dislocated body parts in good enough condition to be donated following an accident. If a doll had hurt herself, my sister and I would take it to my dad, who would consult the box and attend to the injured party.

But the highest praise we could heap on my dad – in our opinion – was to tell people: "He can *even* make bunk beds!" The "even" was important because underneath the bunk bed pinnacle lay a host of other practical skills, which were far less interesting to us. It might take days, but he'd fix the washing machine himself. He took our computer apart once, and made it behave as if it were new again. I can't recall a single thing that my dad – an engineer by trade, a physicist by nature – couldn't fix.

My own washing machine broke a few years ago. Determined to get it fixed, I made a few calls. Eventually, someone agreed to come to see it. You can probably guess the outcome: it was quicker and cheaper to buy a new model than mend the existing one.

I don't know how to fix a washing machine, but I know that it's possible. That it takes skill, time and effort. But when we buy stuff, be it a washing machine or any other consumer product, how often do we think about the people behind its construction? Of the resources used to make it? With so many products that are either built for obsolescence or for trend cycles – and sold for prices that don't reflect the true cost of their existence – is it any wonder we give them so little value? And buy new without a second thought?

According to Friends of the Earth, the world's richest countries consume on average ten times as many materials as the poorest, with North America and Europe boasting the biggest material footprints on the planet. The UK is hugely dependent on other countries' minerals, raw materials, water and land. If everyone lived like the average US citizen, we'd need around four Earths to sustain ourselves.[4] From digital gadgets and personal care products, to clothes and household appliances, we choose to live in a way that our planet cannot sustain.

In his book *The Day the World Stops Shopping*, author JB MacKinnon conducted a thought-experiment by asking the question: can we solve the consumer dilemma? When I spoke to JB, he told me that one of the most important voices in his book is that of Abdullah al Maher, the CEO of a knitwear manufacturer in Bangladesh, where over a third of manufacturing jobs and nearly 85 per cent of exports come from the apparel industry. You'd think Abdullah would oppose the idea of reducing global consumption. Quite the opposite. He makes an impassioned plea for us to ask: Who made our clothes? How did they get to me? How were the materials grown? How are they so cheap?

The 21 essays in this book vary considerably in subject matter, but many – if not all – are united by a common thread: the things we don't see. Hidden, obscured, invisible. These words are peppered throughout *Three Things to Help Heal the Planet* and highlight the crux of the problem within a consumer society: disconnection. We interact with purchases and experiences only from the point of transaction; we have little connection to the natural world and to the hidden people who bring them to us.

But it is all connected.

In her essay "Grow Food, Dismantle the System", the writer and organic food grower Claire Ratinon recalls the moment when she first witnessed food being grown on a rooftop in New York. "Until that Saturday afternoon", she writes, "I'd only ever interacted with food when it was on my plate. It was there whenever I was hungry."

Last summer, I lazily placed a butternut squash plant into a flower bed in my back garden. Before I knew it, that tiny plant had multiplied – thanks to some busy pollinators – to dominate a 5 x 1 metre stretch of soil. Pops of yolk-yellow flowers blossomed in abundance. Tiny squashes shaped like bottom-heavy hourglass timers began to flourish. Dozens and dozens, jostling with the flowers I couldn't name. I pictured the salads and soups I would make. Only one survived.

The soil, the sun, the pollinators – they played their part. But I didn't give those squashes the care and attention they needed to grow. It's only when we engage with the process ourselves, or closely watch the work of others, that we appreciate the food we so greedily eat, beyond what we take from the shelves in a shop.

Our disconnection is further exemplified by our relationship with the personal care industry. In her essay, the make-up artist Khandiz Joni discusses our singular approach to beauty purchases: we worry about the effects of chemicals on our skin yet rarely consider their social and environmental impact. "Cosmetics that go onto the shelves have to be tested; there is legislation to protect us", Khandiz writes. But who protects the oceans from the invisible microplastics released into our waterways when we rinse the silicone-filled conditioners from our hair?

Hidden, obscured, invisible. The persistent presence of these words vividly conjure the opposite through its omission: transparency. Do we know what's in our shampoo? Do we know who makes our clothes? Do we know where our butternut squash comes from? Perhaps that's a good rule of thumb: only purchase when we have the answers.

1

JB MacKinnon

The Day the World Stops Shopping

I was raised in a household where issues of the day were served up alongside the food. We would eat, and we would argue. Every night, at the table with my parents and three brothers, we contested the values we held. You didn't have to present an opinion, but if you did you had to be prepared to defend it. I grew up curious about the question of how best to live.

The thread that pulled that curiosity together – and continues to pull at it now – is the changing human relationship with the natural world. At the turn of this new millennium, according to the United Nations, consumption quietly surpassed population as our greatest environmental challenge. When it comes to climate change, species extinction, water depletion, toxic pollution, deforestation and other crises, how much each one of us consumes now often matters more than how many of us there are.

We all saw – we all felt – the dynamics of that relationship in fast-forward during the coronavirus pandemic, which initially triggered one of the sharpest declines in consumer spending ever recorded. Shuttered shopping districts, empty airports, boarded-up restaurants, millions of people out of work. Equally undeniable, though, were the shockingly blue skies over Los Angeles and London, the sea turtles and crocodiles taking over tropical beaches normally invaded by mass tourism, the fresh air in Beijing and Delhi, the steepest drop in greenhouse gas pollution on record (in some nations, carbon pollution fell by as much as one quarter).

Since World War II, global carbon dioxide pollution has fallen only four times: in the mid 1980s, the early 1990s, 2009 and 2020. In none of these cases was the decline the result of decoupling, green growth or any other purposeful action to protect the planet; each involved severe and widespread economic downturns. Emissions fall when the world stops shopping.

I call this the consumer dilemma: the planet needs us to slow our consumption; the economy needs us to consume more and more. Before the pandemic, I'd embarked on a thought experiment. What would happen if the world stopped shopping? Specifically, and for simplicity, what would happen if global consumer spending dropped by 25 per cent? At the time, many experts I spoke to refused to even entertain the figure. But in early 2020, when US household spending dropped by almost 20 per cent, retail sales in China fell by one fifth and personal consumption tumbled by nearly a third in many European countries, the figure no longer seemed outlandish.

My book *The Day the World Stops Shopping* is the result of that thought experiment. Can we solve the consumer dilemma? Yes, we can. But what I've learned is that a reduction of 25 per cent overnight is not the answer. By reducing consumption at that level, we see the immediate release of human pressure on the natural world. We also see tremendous economic challenges for many. We've been talking too simply about simple living: it can be deeply satisfying to us as individuals, but if too many of us do it, it creates serious problems. For all of us to live more simply, we have to create a different kind of society and a different kind of economy.

One of the most important voices in my book is that of Abdullah al Maher, the CEO of Fakir Fashion, a knitwear manufacturer for global brands such as H&M, Zara and C&A. In Bangladesh, where one fifth of residents live below the national poverty line, the garment industry provides jobs to more than four million people. Six out of ten of them are women. Maher's company manufactures a mind-boggling 200,000 articles of clothing every day. When I asked him about reducing global consumption, I assumed he would oppose the idea. Instead, he answered in the tone of one sharing a secret. "You know," he began, "it wouldn't be so bad."

There's an old saying: if something's too cheap, someone else is paying. "When you buy a fast-fashion T-shirt for four dollars, you never ask, 'How could this shirt have landed in Berlin or London or Montreal for this price? How does the cotton get grown, ginned, spun, woven, dyed, printed, sewn, packed, shipped, all for four dollars?'" he told me. "You've never realized how many lives you are touching, all because *your* payment doesn't pay for their wages."

Maher thinks we need to transition away from cheap, disposable garments and instead pay a fair price for fewer well-made clothes – even though it will be a costly and difficult transition. It makes more sense to move painfully in a direction that ends up in a better place, than to continue moving down the brutal path we're currently on. "There'll be no rat race then," Maher said. "There'll be a real race."

More than a decade ago, I had an experience that reshaped the way I think about such changes. In 2007, my partner and I wrote a book called *The 100-Mile Diet*, for which for one year we ate only food and drink produced within a 100-mile radius of our home in Vancouver. It wasn't easy, but it was fun, interesting and completely transformed the way we eat. We still eat about 90 per cent locally now because we prefer it. It tastes better. We feel healthier. We're more connected to the people and places that produce our food. You couldn't pay me to go back to the global supermarket.

If this sounds like an argument for making change through individual lifestyle choices, it isn't – at least, that's not the whole story. Our book encouraged a lot of people to eat locally, yes. But that shift was supported by a strong local food movement. Through the combination of individual and collective action, the agricultural landscape was transformed. People were able to make a living doing smallholder farming. The growing of grain on the coast of North America started up again. The range of crops expanded. We once didn't have winter farmers' markets in Vancouver; now we have them weekly. Far more restaurants are using local food.

When you see the combination of individual action and system change play out, it's tremendous. And that was the big learning. People will only stick to changes if two things happen: first, if it feels like improvement of their quality of life instead of sacrifice; second, if systemic change supports personal action.

We all know how to contribute as individuals through lifestyle choices, and it's a really good point of engagement with issues. But how can we effect systemic change? You could write to your favourite brand and ask that they make more durable goods instead of relying on high-volume sales of disposable goods. Whether you work or run your own business, you can influence a shift toward practices that will result in lower consumption. If you're an economics student, you can help draft policies to support a lower-consuming society. If you study art, you can show the beauty of worn and weathered objects. As a voter, you can support politicians who will fight income inequality – a major driver of consumer culture.

Done collectively, these actions help to influence legislation: the intersection between individual and systemic change. Right-to-repair laws, lifespan labelling on products, reduced taxation to support re-use and repair – these are examples of concrete steps society can take to combat planned obsolescence of goods and make it easier for us all to buy better.

Sometimes it feels like we're asking, "How can I change without changing anything?" We can't. We do have to confront certain realities and be prepared to make real changes. My focus as a consumer now is fewer, better everything – fewer goods, fewer services, fewer consumer experiences, fewer hours spent on social media.

A lower-consuming society isn't a society that doesn't consume. Ironically, one of the qualities of a lower-consuming society might be a stronger relationship to material goods: a life in which you still have possessions, but long-lasting ones that you keep for years if not a lifetime. A life where you still travel, but less often, and in a more engaged way. A life in which you feel relief from many of the social pressures around income and status we feel today. A life where you have time to invest in what matters. A life with stronger connections to people you care about, to the natural world. A life where you engage with issues that are bigger than yourself, where you make things, where you develop an authentic identity. This is – I think – a better way to live.

But it can be hard to imagine. Many of us come from a place where we mark life's moments with consumption. We show our love through consumption, through gift giving. We form our identity through consumption. We map our way forward through consumption. Will we really enjoy a simpler life?

The psychologist Tim Kasser, who has studied materialism for over thirty years, sees evidence we will. He points out that people who practise simpler living today come to it through many different doorways. Some want to reduce their debt. Others are overworked and want some kind of life balance back. Some choose simplicity for environmental reasons. Despite these differences, they tend to pursue simpler living in the same ways – and often dive in deeper than they expected. Once we try simplicity, we often find we want even more of it.

What is it like to live in a world of less? Here's one more tantalizing glimpse. Early in the pandemic, mass tourism to Hawai'i ended. Biologists reported dramatic changes on coral reefs that had been heavily visited. They saw monk seals, dolphins, sea turtles and schools of big fish return. But they also saw a change in the animals' behaviour.

There's a continuum. On a reef where humans hunt fish, they are naturally skittish and fearful – you won't see very many. On a reef where they're not hunted but are exposed to people all the time, many sea creatures disappear; the rest become blasé. A reef where a human is neither a threat nor a constant presence is best of all. It's there that nonhuman life is neither fearful nor indifferent, but curious enough to approach us. In a lower-consuming world, when we do make our journeys to wild places, we stand a better chance of a bit of magic – to connect across the void between species, two beings face to face, taking an interest in each other.

2

Khandiz Joni

The Bathroom Epiphany

I grew up on film sets in Cape Town, watching my mother work on TV commercials in the art department. The hours were long, the work was hard and for a department using the title "art", it wasn't very creative. As an art student, I wondered how I was ever going to make a living from being an artist. One day, while visiting my mother on set, I saw the make-up artist at work. Not only was she doing something that looked like art, but apparently she was also better paid! I didn't become a make-up artist because I loved beauty products. For me, it was a painting medium with an income.

But 12 years later, when I settled in the UK, something happened. I remember the moment as clear as day, standing between the kitchen and stairwell in my shared London apartment. It dawned on me that I lived my life one way – naively compassionate about my impact on the planet – but the work I did simply didn't align with that. Being a workaholic (as a freelancer, you never know when your next paycheck will arrive, so you tend to work non-stop) meant spending the majority of my time working in an industry, and with materials, that perpetuated things I completely disagreed with. It was this realization, in that moment, that ignited something in me and has been driving me forward every day since.

My journey to what I call an eco-ethical artist, and the subsequent use of conscious beauty products, began from an environmental standpoint, rather than from the health perspective of many of my peers. There was the obvious – plastic pollution from all the packaging – but what really piqued my interest were the wider social and environmental benefits of organically farmed ingredients that were being used in these novel products.

Some in the green beauty world will talk about why you should be terrified of certain products because they contain a carcinogen or endocrine disruptor. Or they'll use misleading language such as "chemical free", which is completely inaccurate and, in my opinion, as bad as greenwashing (when corporations falsely brand their products or actions as eco-friendly, green or sustainable). While it's true that many conventional beauty and personal care products contain contentious ingredients that can negatively impact human health, the reality is that the quantity found in a single bottle of shampoo or a jar of our favourite moisturiser won't harm us. Cosmetics that go onto the shelves have to be tested; there is legislation to protect us.

What is less often spoken about is the bio-accumulation of certain chemicals in our bodies, and in the environment. For example, the silicone in hair conditioners is an inert ingredient, which really just means that it's cheap and functional (by providing a bit of slip and slide). It doesn't harm the skin, but it certainly doesn't benefit it either. What it does do, however, when we wash our hair and rinse away the product, is send microplastics into our rivers and oceans – the kind of plastic that beach cleans can't resolve. As my favourite Tanzanian proverb goes: "Little by little, a little makes a lot." I think about this sentiment often. The real problem is that we think in the singular, rather than the wider, connected implications of our choices.

When we talk about the climate and ecological emergency, we need to remember that we are talking about a social emergency. The products we use to primp and prime ourselves are directly connected to climate-altering emissions, resource scarcity and biodiversity loss. By choosing to invest our hard-earned money into brands that are actively reducing their climate-altering emissions, water consumption, waste output and reliance on virgin materials from extractive industries, we can make a difference.

Consider that the most prolific cosmetic ingredient is water. Water is going to be the next thing that world wars are fought over. Question the use of natural ingredients; natural doesn't always mean sustainable. Is it acceptable to extract exotic ingredients from remote locations, from Indigenous communities that rely on them for food and medicine without ensuring they benefit from the harvesting of those ingredients? Should we take frankincense from sacred trees – in short supply and high demand – for the sake of our skincare routine, or might a synthetic version created using green chemistry to synthesize them be more sustainable, despite the synthetic label? If you care about how women are treated, if you care about Black lives, have you considered where and how your favourite plant waxes are harvested? Ingredients such as shea butter and cocoa butter, which you find in everything from body wash to lip balm, come from countries such as Kenya or Ghana, and are harvested by a majority-female workforce, who, unless they are part of a Fairtrade initiative, are often exploited. This, in turn, can lead to the exploitation of their own natural resources – the same trees that are cut down for burning are precious commodities for these communities – just in order to survive. Or what about exploitative child labour conditions in the mica mines of India so we can have sparkly bronzers and shimmery eyeshadows? Choose products from brands that share your ethics, and can back up their claims.

So in that one moment, in 2012, standing in my apartment, I made a choice. I decided to systematically finish all the conventional products in my kit, all the stuff I'd been told for years needed to be on my table to do my job, the cosmetics with brand reputations, and replace them with products that aligned with what I stood for. It took two years.

I became a tadpole in a sea of make-up artists. But that difference was a true reflection of who I was. It made me better at what I did because I was no longer reliant on a product – I actually got to use my skills. Plus, being a make-up artist gives you a rare opportunity to be in an intimate environment with somebody else, to talk about the big issues. I've seen people's reactions when I use products bursting with vitality, ones that smell different ... wholesome. I can see it makes them sit up and take notice, and it makes them feel comfortable in their skin.

From there, it's a jumping-off point to talk about the beauty industry's impact on the environment, on supply chains, on people and communities. Naturally, there were some challenges in giving up the conventional brands, but I just turned work down that I felt I couldn't achieve with a slimmed-down eco-ethical kit. I know that's not possible for everybody – I'm not a parent, for example, so I could afford to take a financial hit, to live more frugally until I got the work that aligned with my values. The bottom line is, everybody can do something.

As a starting point, that something could be a bathroom audit. Start by taking all your personal care products, from make-up to moisturizers, and place them into category piles: shampoos, conditioners, shower gels, eyeshadows. For any product that has expired, unless it's medical, follow your senses – smell it, feel the texture. A product may have a six-month expiry date, but a lipstick can last for years. Once you've put the products into categories, you can see what you need to finish off first, and you can see what you do and don't use. It's an exercise that goes beyond the environment to the financial: are you buying stuff you don't need and could you save money as a result?

Once you've identified the products you actually use, consider what you'll replace them with only once they're finished. Remember: the most sustainable product is the one you already own. To do that, align your purchasing behaviour with your personal values. If you're vegan, ask if the brand cares not only about the ingredients but also for the environments in which animals live. Unless it's a natural vegan brand that uses regenerative ingredients, the items will be by-products of the petrochemical industry, which is not only destructive because of its emissions but also harms animals' habitats.

Sunscreen is another example. Lots of brands claim their products are reef-safe. This is not a technical term, and often just means that the sunscreen excludes certain proven problematic ingredients such as oxybenzone. But unless the remaining ingredients have been tested in how they work together, and how they react with salt water –

another chemical – and what impact they have on the marine ecosystem, brands cannot make that claim.

Fortunately, in 2020 the Competition and Markets Authority began a review of misleading green claims after launching a new set of guidelines, and began imposing fines in late 2021, with the beauty industry among the sectors facing scrutiny. But we have to keep the pressure up; we have to ask the big, sometimes uncomfortable questions, and to look beyond the obvious. In my experience, aside from cost and efficacy, there are six value streams that influence our beauty purchases: environmental, health, animal welfare, and social, religious or cultural beliefs. By starting with what matters most to us, we automatically make the time to investigate more than just skin deep.

The truth is, there is no magic pill, but certifications are a good place to start. COSMOS certification brings together organizations such as The Soil Association and Ecocert, and certifies products based on their organic or natural ingredients, and sustainable production standards. By choosing an organic-certified product, you take care of a lot of concerns.

I've been a make-up artist for two decades now and, since having that epiphany ten years ago, I've had another. Again, I remember the moment with absolute clarity, standing on a beach in Hastings in November 2019, unaware that a global pandemic was on its way. I decided to no longer be a commercial make-up artist. As wonderful and beneficial as eco and ethical products are, there is still too much waste. And there are so many incredibly talented make-up artists with an emotional connection to the product that I don't need to fill that space anymore. Instead, I've retrained and qualified as a sustainability advisor. From now on, I'll be combining that with my years of experience to advise the beauty and creative media sectors from within.

I will still paint – that will never go away. I will keep all my brushes, all my tools. I will do your make-up, if you like, with your products. But I don't need suitcases of eyeshadows to do what I love. The question is, how do we do the most with the least amount of stuff? And that is what excites me now.

3

Claire Ratinon

Grow Food, Dismantle the System

Beautiful scars ran down the bizarre, bulbous shapes in glorious shades of peach, purple and black. Sweet and complicated, these gnarly creatures tasted of sunshine. The sign on the door had said "Come and visit our rooftop farm" – this was an eight-storey building in Queens, New York, and implausible images of fields filled with crops came to mind. Not these precious heirloom tomatoes. Not peppers, not kale, not basil, growing in abundance in an acre of productive land overlooking the city. It was a life-changing moment as I realized that I had never seen food being grown before.

I had a suburban upbringing. Success meant the city, a professional job, steady income, status. I'd been in New

York for two years working as a documentary filmmaker and, before that day, never thought about who grew my food. It never occurred to me that there was a whole system in place that fed me. Until that Saturday afternoon, ten years ago, when I found myself at Brooklyn Grange Farm by chance, I'd only ever interacted with food when it was on my plate. It was there whenever I was hungry.

And yet I loved food. I was obsessed with it; my whole family was – at taste level. I wanted to participate in the magic of what I was seeing, so I started volunteering at Brooklyn Grange every Saturday. I sowed seeds and watched them grow. I witnessed the volume of labour that goes into harvesting, packing and transportation. By working in a productive space, you see the many, many layers of what it takes for food, from its inception, to reach your plate.

In turn, you also see that much of this process is systematically obscured. So many actively engaged human hands participate in the systems of growing, producing, processing, packaging and delivering our food until it is exchanged for money – but we don't see these hands. These processes are obscured on purpose. They're obscured and it causes the people who do that work to be devalued. They're obscured and it causes the natural world to be devalued. Learning how to grow food enabled me to see the exploitation behind this curtain. In the food system first, but then in everything else.

We don't necessarily have to understand every section of the production line that delivered that bowl of pasta to us. But once you start pulling at that thread, if you can use it as a lens to see the level of complexity that goes into it, the least you can do is pause in gratitude before you eat every day. We can use that understanding to make choices that are in the best interests of the natural world and the people who are part of the processes that feed us. That is what learning to grow food can offer us.

Growing your own food is never going to feed you in a meaningful way, unless you have great swathes of land and people working for you. It's hard work and it's important we don't devalue that with fallacies of self-sufficiency. The point of growing our own food – for me – is to re-embed ourselves into understanding what it takes to do so, to put centre those who do the incredible work of feeding us every day with reverence and respect. I've only ever grown food on a relatively small scale, but I know what it takes to feed people. The sheer volume of effort, of human effort, ecosystem effort, creature effort, plant effort – all of that energy convalescing perfectly in order for us to eat. It is a miracle that it happens with such abundance.

So how could you start to grow your own food at home? Firstly, do you have outside space? Do you have access to soil? Do you have enough sunshine? A north-facing garden, however large, is frustrating to the point of impossible. So if you don't have these things, find somewhere that does. When I moved back to London from New York, transformed by those two seasons at Brooklyn Grange, the windowsills in my tiny, dark flat were too small to hold plant pots. So I started looking for growing projects and gardens in east London. Most growers will welcome your help, they rarely turn away an extra pair of hands.

If you are lucky enough to have access to sunshine at home, I suggest trying your hand at growing edible plants in a pot. If all you have is space for a two-litre pot on a sunny doorstep, all you need is some compost and seeds to watch something grow.

My heart belongs to leafy greens. When I moved back to London and joined the social enterprise Growing

Communities, these were the focus of their market gardens. So to new growers, I say, start with salad leaves. With beautiful lettuces, brassicas, herbs. Not only are they the easiest edibles to grow, but if it's not going very well, you can still eat them. They don't need quite as much sunshine, nor as big a container as other vegetables because they don't have large root systems.

You won't be self-sufficient, not even in salad leaves, but you will get to be part of this humbling and magical process. It's one of the acts that truly unites us all. Our common ancestors grew their own food, and it's still happening all the time, we rely on it entirely. It's complex and refined and skilled, and if you can go into it with that sense of meaning in your heart, even if it doesn't work at first – which only adds to the wonder and learning – you will witness one of nature's incredible processes, and be responsible for that harvest.

Those who grow our food, who make it possible for us to eat, should be at the centre of our dinner table. Before we put fork to mouth, let us pause and say thank you.

Three Things to Do

When my shampoo ran out, I decided to follow Khandiz Joni's advice and buy an organic alternative to the brand I normally use. It smelled delicious and the packaging was recycled, recyclable and refillable. But when I washed my hair, there was little lather. The wet strands didn't feel as clean as with my old brand. I decided to dry my hair to see if that would make a difference. The result was better but still not as grease-free as I'd like. I was so disappointed, so frustrated. I really wanted the organic version to work. So I read up on the brand and found some advice: give a new product some time, allow your hair and scalp to adjust. After about six washes, it did.

Personal care products are just that: personal. We develop a close affinity with certain brands so it can be hard to part with them. You could try writing to your favourite brands if their ingredients list isn't up to scratch. Or you could give new products a try and make them your new favourites.

ACTION 1: CONSUMERISM

To do: buy fewer, better everything – fewer goods, fewer services, fewer consumer experiences.

- Before you buy something new, ask yourself: do you really need it?
- If you do, write to your favourite brand to enquire about their values and processes – and if they don't align with yours, ask that they make their products more durable and ethical.
- Do you devote too much time, money and head-space to consumer experiences – from shopping to social media – that don't make you happy? Could you minimize that interaction and replace it with activities that you love?

ACTION 2: PERSONAL CARE

To do: conduct a bathroom audit. Place all personal care products – from make-up and moisturizers, to shampoo and hair gels – into category piles to determine what you do and don't use. Once it's time to replace a product, do so with a brand that cares about its impact on people and the environment.

- Before you buy more personal care products, check the contents of your bathroom – is there a shampoo bottle hidden at the back of the cupboard that you could use?

- Are you buying things you don't need, so could you save money by striking them off your shopping list?
- When you do need to replace a product, check the ingredients and certifications: do you prefer organic products? Are you vegan? Make your next purchase count.

ACTION 3: GROWING FOOD

To do: grow some of your own food. You won't be self-sufficient, but you'll learn and appreciate the skill and complexity of growing food – and be responsible for that harvest.

- Do you have a south-facing windowsill, patio or garden? If so (and you've never grown food before), could you start by growing leafy greens in a two-litre pot on a sunny spot?
- If you can't grow your own food, could you volunteer with a local growing community group?
- Next time you buy fruit and vegetables, check who grew them and where they were grown. Could you purchase in-season produce from local suppliers and communities?

EAT

- The global food system is responsible for 30 per cent of all greenhouse gas emissions.[1] It is the human activity with the single biggest impact on the planet.[2]
- Thirty per cent of the food we produce is wasted – about 1.8 billion tonnes a year. If, as a planet, we stopped wasting food altogether, we'd eliminate 8 per cent of our total emissions.[3]
- The average global citizen has a CO_2e[4] footprint of approximately 4.5 metric tons per year, but it must not exceed 2.1 in order to meet the Paris Agreement's goal of limiting global temperature increase to 2°C. Not eating animal products for breakfast and lunch saves 1.3 metric tons per year.

"If you want to change the world, you have to throw a better party." It was when I first heard those words, more than three years ago, that I knew an idea could become the book you are reading today. They were spoken by the anti-food-waste campaigner Tristram Stuart, the first contributor to *Three Things to Help Heal the Planet*. He generously gave me hours of his time to explain the relationship between food production and the climate crisis.

In his articulate and passionate explanations, Tristram's anger is often tempered by his belief in the power of food to bring people together. His favourite word is "companion" and one of his most recent campaign vehicles is beer.

Tristram is the founder of Toast Ale, a beer launched in the UK in 2016 made from fresh surplus bread instead of barley, using less land, water and energy in the brewing process. Since inception, the company has saved 42 tons of CO_2 emissions and 257,413 litres of water. It certainly offers a different definition of responsible drinking.

Three years earlier, and together with Thomasina Miers, the co-founder of Mexican restaurant group Wahaca, Tristram launched The Pig Idea, a campaign calling for change in

European law to the way we feed pigs, by returning to the traditional practice of feeding them surplus food waste rather than crops that people could otherwise eat, such as wheat, soy and maize. The campaign began with the rearing of eight pigs at Stepney Farm in east London and culminated in a party at Trafalgar Square, where some of the UK's best-known chefs served up their favourite pork dishes to thousands of members of the public – for free.

Pigs were the reason Tristram began campaigning in the first place. When he was 15 years old, growing up on a farm in Sussex, he bought some pigs and started feeding them in the most traditional and environmentally friendly way. He'd ask his school canteen for the day's scraps. He'd pick up day-old bread from the local baker. He took some unwanted potatoes from a local farmer, who would have otherwise thrown them away because they were the wrong shape for the supermarkets. One morning, he noticed amongst the "scraps" a tasty-looking sundried tomato loaf. He picked it up, sat down and ate breakfast with his pigs. Companionship is not limited to humans.

In the UK alone, households throw away 4.5 million tonnes of food each year that could have been eaten, the equivalent of eight meals per household each week.[5] Tristram's discovery that much of the food he was feeding his pigs was fit for human consumption led to his first large-scale party – also at Trafalgar Square – which became known as "Feeding the 5,000" and kicked off a global movement against food waste.

Tristram's view that environmentalism needn't always be about sacrifice is shared by Melissa Hemsley, a chef, Fairtrade Ambassador and self-confessed lover of leftovers. In the foreword to her book *Eat Green*, the farmer and founder of Riverford, Guy Singh-Watson, writes: "I first met Melissa foraging for samphire on a salt marsh near our farm in Devon. As a grumpy, old, welly-wearing farmer, I am programmed to rebuff the arrival of fashion and lipstick in my fields, but even I have to acknowledge that [...] responsible eating needs to be enjoyable, easy and accessible if it is to become mainstream, as it must."

Melissa bears a huge sense of self-imposed responsibility, but she confronts it with contagious joy. My interviews with her were some of the longest I've conducted, such is her passion for making change fun. She has a knack for naming her creations – Clear-The-Fridge Curry and

Fridge-Raid Frittata are two of my favourites. In her honour, I now make Melissa Minestrones from whatever veg is left in the fridge.

According to the United Nations Food and Agriculture Organization, livestock is the leading cause of climate change, responsible for 14.5 per cent of annual global emissions.[6] A report by Friends of the Earth and the European political foundation Heinrich Böll Stiftung found that, between 2015 and 2020, global meat and dairy companies received more than $478bn in backing from 2,500 investment firms, banks, and pension funds, most of them based in North America or Europe. The 35 largest meat and dairy companies emit more greenhouse gases than the economy of Germany.[7]

Meat forms the basis for Jonathan Safran Foer's argument in his essay, an edited extract from his brilliant book *We Are the Weather: Saving the Planet Begins at Breakfast*. His advice may be simple – and unpopular with many – but it's an articulate delivery of science and stories. At no point does Jonathan preach; in fact, he often admits to personal mistakes, including eating burgers at airports for comfort.

"When a radical change is needed, many argue that it is impossible for individual actions to incite it, so it's futile for anyone to try", Jonathan writes in *We Are The Weather*. "This is exactly the opposite of the truth: the impotence of individual action is a reason for everyone to try."[8]

4

Tristram Stuart

Throw a Better Party

My favourite word is companion. I like to unpack its etymology: *com* means "with" and *pan* means "bread". A companion is someone you share food with, and sharing food as a way of building friendship is universal. When you think about the 1 billion tonnes of food thrown away each year, it's not just all the wasted carbon emissions, land, water and labour; it's also a failure of good companionship on a global scale. Think of all the friends we could have made if we'd shared that food rather than thrown it away.

Climate change deniers can argue about the maths, but no one can deny that humans replace wild forests, wetlands and grasslands with fields for food production. This is happening on a colossal scale because of the amount of food we eat – and throw away. The result is species extinction, soil erosion and a massive contribution to greenhouse gas emissions: food production is the human activity with the single biggest impact on our planet. And so much of that food is wasted. If it were a country, food waste would be the third biggest emitter of greenhouse gases, after the US and China.

The bright side is that food is an incredibly powerful, cultural tool. We come together around food like nothing else. When you bring that together with our biggest asset – an innate desire to be connected – you could transform what's currently an extremely damaging system into potentially the single biggest tool for tackling all of the big environmental problems.

Many people think environmentalism is about sacrifice. I try to live by the maxim that if you want to change the world, you have to throw a better party. After my book *Waste: Uncovering the Global Food Scandal* was published in 2009, we decided to invite 5,000 people to Trafalgar Square and feed them free, delicious food made from ingredients that didn't meet supermarkets' cosmetic standards. I didn't know if anyone would come. On the day itself, it began to snow. As I stood on the balcony overlooking Trafalgar Square, I saw huge crowds gathering. People came, we fed them, everyone had fun.

"Feeding the 5,000" kicked off a global movement against food waste, creating new laws, changing the big supermarkets' approach to wasting food and persuading British people to cut food waste in their homes by nearly one third. We all have the power to co-create a different food system. Starting the journey by simply buying less and eating everything we buy can be empowering: this alone can help decrease your contribution to the environmental crisis. After that, we can think about next steps. It'll be different for every individual.

Perhaps you'll halve your meat and dairy consumption, and halve it again a couple of years later. Perhaps you'll ask local stores if they donate unsold food to charities that feed hungry people, and boycott them if they don't (you can always check their rubbish bins round the back). Perhaps you'll volunteer for a local food redistribution organization or pressure group, or photograph egregious instances of food waste and publish them, naming and shaming the companies. Or maybe you'll just make sure that next time you have a leftover pizza slice, you'll offer it to a friend, or even better a stranger, rather than dump it in the bin – and that act of companionship will make you a new friend. The single biggest thing we can all do is love nature: to treat the Earth like a cherished lover whom we depend on and want to sustain. From this flows all the other things we need to do to avert disaster.

We vote on the food system every day with the money we spend. We are driven to make certain choices by incredibly powerful forces and marketing budgets of companies such as Coca Cola, whose core product is laced with a semi-addictive drug (caffeine) and contains more sugar than is good for a child to consume. We have a lot stacked against us. So for some people, eating their leftovers is achievable, but for many of us, going further is our responsibility. Governments are not taking a fraction of the action required to avert this impending catastrophe that will affect all life on Earth. They need a mandate to do that. I intend to live a good life while fighting the good fight. How far will you go?

5

Melissa Hemsley

In It Together

We all do it, don't we? Question our expertise, our credibility. Surely there is someone better qualified than me to stand up and tackle the important things? I remember when "The Sustainability Sessions" (a community of live events to help us make sustainable choices) was just the nugget of an idea. My exasperation – *I can't possibly talk about these things until I know more!* – countered by the calm of my managers: "Don't you think people would like to see you learning?" And expose my ignorance? My vulnerability? But I quickly realized that most of us were – are – still learning. There is no "perfectly sustainable", no quick fixes – we're all on individual sustainability journeys. So, show that it's relatively new to me; show my stupid questions; show where I'm getting the answers from and learn with me. That's how we make the discussion relatable, achievable and inclusive.

I don't need to be an environmental expert to help make the world a better place. Neither do you.

When it comes to food, a major reason that stops us from making fundamental changes is our increasing disconnection to food and nature. How can people tackle the fact that food production is the biggest contributor to climate change when they're surprised that you can easily swap ingredients in a recipe? That the mileage on a butternut squash in May is far greater than in November? We need to get better connected to our farmers, the seasons and "state of our plate".

But because we don't feel connected to the ingredients we cook with, we lack confidence – and waste food. The way I see it, wasting food is a waste of time, resources and free flavour. I get so much satisfaction from showing people how to adapt recipes to their tastes, energy and kitchens – and seeing their reactions. They always say: "Oh, but it's so easy! How cool that I can swap this vegetable for something I need to use up!" Cooking a meal together builds confidence. Showing people that wasting much less is do-able to the point of pleasure makes the habit. That, in turn, is infectious.

So how do you start? Put simply, take half an hour to create a weekly meal plan. You'll shop more efficiently, cook more efficiently and waste less. You can do this within the context of what you already buy to make your food (and money) go further, or you can make significant changes.

As I write, I'm in my kitchen: I've got my fresh, freezer (I depend on my freezer) and cupboard food. Where do you spend your money? Look at an old shopping receipt, at the contents of your food bin. Can you commit to spending one third of your shopping budget with businesses that are local, organic, Fairtrade, ethically sourced or independent? We can vote for a fairer, better and more sustainable food system by using our spending power. If possible, can you commit to a regular food delivery box, filled with seasonal produce and no plastic packaging? Check out the fantastic database Farms To Feed Us or ask someone who has done the hard work already. You save time and energy by making the right choice at base level.

Consult your household when planning the weekly meals. Partners, children, flatmates – I remember watching those multiple tubs of hummus and yoghurt go off when I lived with friends. Share! Make it fun! So, we're definitely having Bolognese this week because it's your favourite. Good quality organic mince will cost about £1.50 more per pack. Could I use 80 per cent meat and 20 per cent lentils? Grate in some courgettes and extra carrots, to get much more veg in and stretch the meat further? Put the rest in the freezer? You're diverting your money to better-quality meat, but eating less of it and more veg.

As you reach the end of the week and your fresh food is looking a bit limp, embrace using it up. Try a veg and herb-packed Fridge-Raid Frittata or Friday Fritter (depending on how many eggs you have). Or a quick Clear-The-Fridge Curry. With ripe bananas, bruised apples and soft pears, make a Fruit Bowl Bake with porridge oats, which you can eat for breakfast. It's on days like these when I hear my mum's words of wisdom: no need to go shopping until the fridge is empty and make your freezer your friend.

6

Jonathan Safran Foer

We are the Weather: Saving the Planet Begins at Breakfast[6]

Conversations about meat, dairy, and eggs make people defensive. They make people annoyed. No one who isn't a vegan is eager to go there, and the eagerness of vegans can be a further turnoff. But we have no hope of tackling climate change if we can't speak honestly about what is causing it, as well as our potential, and our limits, to change in response. So I'll name it now: we cannot save the planet unless we significantly reduce our consumption of animal products.

My book, *We Are The Weather*, is an argument for a collective act to eat differently – specifically, no animal products before dinner. That is a difficult argument to make, both because the topic is so fraught and because of the sacrifice involved. Changing the way we eat is simple compared with converting the world's power grid, or overcoming the influence of powerful lobbyists to pass carbon-tax legislation, or ratifying a significant international treaty on greenhouse gas emissions – but it isn't simple.

In my early thirties, I spent three years researching factory farming and wrote a book-length rejection of it called *Eating Animals*. I then spent nearly two years giving hundreds of readings, lectures and interviews on the subject, making the case that factory-farmed meat should not be eaten. So it would be far easier for me not to mention that in difficult periods over the past couple of years I ate meat a number of times. Usually burgers. Often at airports. Which is to say, meat from precisely the kinds of farms I argued most strongly against. And my reason for doing so makes my hypocrisy even more pathetic: they brought me comfort. I can imagine this confession eliciting some ironic comments and eye-rolling, and some giddy accusations of fraudulence. How could I argue for radical change, how could I raise my children as vegetarians, while eating meat *for comfort*?

We do not simply feed our bellies, and we do not simply modify our appetites in response to principles. We eat to satisfy primitive cravings, to forge and express ourselves, to realize community. We eat with our mouths and stomachs, but also with our minds and hearts.

There is a place at which one's personal business and the business of being one of seven billion earthlings intersect. Climate change is not a jigsaw puzzle on the coffee table, which can be returned to when the schedule allows and the feeling inspires. It is a house on fire. The longer we fail to take care of it, the harder it becomes to take care of, and because of positive feedback loops – white ice melting to dark water that absorbs more heat; thawing permafrost releasing huge amounts of methane, one of the worst greenhouse gases – we will very soon reach a tipping point of "runaway climate change", when we will be unable to save ourselves, no matter our efforts.

When we think back on moments when history seemed to happen before our eyes – Pearl Harbor, the assassination of John F Kennedy, the fall of the Berlin Wall, September 11 – our reflex is to ask others where they were when it happened.

The word "crisis" derives from the Greek *krisis*, meaning "decision".

The environmental crisis, though a universal experience, doesn't feel like an event that we are a part of. It doesn't feel like an event at all. And despite the trauma of a hurricane, wildfire, famine or extinction, it's unlikely that a weather event will inspire a "Where were you when ..." question for anyone who didn't live through it.

But future generations will almost certainly look back and wonder: where was our selfhood? What decisions did the crisis inspire? Why on earth – why on Earth – did we choose our suicide and their sacrifice?

Perhaps we could plead that the decision wasn't ours to make. Being mere individuals, we didn't have the means to enact consequential change. We didn't run the oil companies. We weren't making government policy. The ability to save ourselves, and save them, was not in our hands.

But that would be a lie.

My book is not a comprehensive explanation of climate change, and it is not a categorical case against eating animal products. It is an exploration of a decision that our planetary crisis requires us to make.

Climate change is the greatest crisis humankind has ever faced and it is a crisis that will always be simultaneously addressed together and faced alone. We cannot keep the kinds of meals we have known and also keep the planet we have known. We must either let some eating habits go or let the planet go.

Where were you when you made your decision?

Three Things to Do

I have four lists on my fridge that I rotate throughout the year. Each contains the names of fruit and vegetables that are currently in season (I copied them from Melissa Hemsley's *Eat Green* cookbook). I take the relevant one with me when I go shopping. Similarly, I now only buy – well, perhaps 95 per cent of the time – Marine Stewardship Council-certified fish. That mainly translates to cod, haddock and some prawns – very rarely to seabass and salmon.

What this means is that I don't always get what I want, but that's not such a bad thing in life. It also encourages me to be creative; sometimes that's rewarding, sometimes I'm still a bit hungry. I've become obsessed with eating leftovers, smelling yoghurt after its use-by date (and continuing to eat it for several days past it) and stewing fruit that's gone soft.

I've never eaten meat because I don't like it. But I like eggs. So I haven't managed to follow Jonathan Safran Foer's advice to the letter. Still, by not eating meat at all, I think I can offset the occasional boiled egg for breakfast.

ACTION 1: REDUCING FOOD WASTE

To do: don't throw food away if it can be eaten. Freeze it or share it – apps such as OLIO connect local people to share uneaten food. And buy less for your own consumption next time.

- Check the contents of your food bin. Do you buy more food than you actually eat?
- Could you ask your local stores to donate unsold food to charities that feed hungry people (and boycott them if they don't)?
- Would you consider volunteering for a local food redistribution organization or pressure group?

ACTION 2: MEAL PLANNING

To do: create a weekly meal plan. By planning your meals a week ahead, you can shop and cook more efficiently, saving money and reducing food waste.

- If you don't already, could you commit to spending one third of your shopping budget with businesses that are local, organic, Fairtrade, ethically sourced or independent?

- Could you commit to a regular food delivery box, filled with seasonal produce and no plastic packaging? Farms To Feed Us is a good place to start.
- Do you really need to go food shopping today, or could you make something from what's still left in the fridge and cupboards?

ACTION 3: MEAT CONSUMPTION

To do: reduce your meat consumption. If you can, go as far as Jonathan Safran Foer's suggestion: don't eat animal products before dinner.

- If you eat a lot of meat, could you halve your consumption over a year?
- Could you halve it again in two years' time?
- Do you know the provenance of the meat you buy? Do your research and buy from local suppliers if possible.

- More than 100 billion pieces of clothing are produced each year,[1] yet 92 million tonnes of textiles waste is are created during the same period. The equivalent to a rubbish truck full of clothes ends up on landfill sites every second.[2]
- The fashion industry contributes to around 10 per cent of global carbon emissions and produces nearly 20 per cent of all waste water.[3]
- Textile production contributes more to climate change than international aviation and shipping combined[4] and creates chemical and plastic pollution.

My Barbie was always the best dressed. Unlike my friends' dolls, she had a stylist: my grandmother. Irene was a seamstress and ran her couture business from home in a converted garage. Growing up in Brazil, my sister and I would mingle with her all-female staff – listening to their gossip, joining in the raucous laughter and hiding under the sewing machines. With the fabric scraps of her clients' dresses, Irene would create extravagant miniature versions for our dolls.

These weren't loose-fitting tunics; I wouldn't be surprised if they were cut from a pattern. Subtle sweetheart necklines and full skirts. Structured, longline ball gowns with frill trims. Bias-cut maxis, all finished with such precision that they'd never fit a Sindy doll. Then there were the fabrics. An emerald silk so shiny I could almost see my reflection. The textured white satin. The contrasting jacquards. The only dress I remember that wasn't made by my gran was a swishy black halter neck. It was also the only one that ever ripped – which Irene mended, of course.

Irene's love of clothes extended to her own wardrobe. Every year, she'd treat herself to a new skirt and my grandad would mark its debut with trademark theatrics.

He'd announce Irene to us, who would slowly walk down the "catwalk" – a marbled staircase to the living room – throwing back her Elnett-sprayed head in a cackle of laughter, her face brilliantly made-up.

I doubt the motivation behind Irene's considered purchases and handmade dolls' clothes was an ethical one; she loved stuff. Take her to a gift shop, and you wouldn't see her for hours. A glass cabinet in her living room was filled with useless ornaments, collected over many years. But she had an appreciation for beauty and an understanding of craft. She cherished what she bought and made; nothing was disposable because she knew what it took to make a well-cut dress.

Many years later, I became a fashion journalist – and accumulated more clothes in a few years than Irene did in a lifetime. Packing up my wardrobe to move house once, I counted 22 pairs of jeans. From high-waisted to flared silhouettes and skinny cuts – in all washes and colours. Some were indistinguishable from each other; you'd need a tape measure to find a difference in the hems. Today, more than 100 billion pieces of clothing are produced each year, with 57 per cent ending up in landfill.

When I began my journalism career as a news reporter, my beats included ethical fashion (an area so tiny I could count my contacts on one hand) and the fast-growing industry darling "value" sector (in other words, the least expensive brands). They were completely at odds with each other. One was about interpreting catwalk trends in the blink of an eye for a fraction of the price. The other was championing Fairtrade principles and organic cotton. One was fast, the other slow. And one was definitely the frontrunner. New dresses, coats, shoes – they were appearing as if out of nowhere, quicker than we could wear them. Luckily, they were cheap as chips!

But someone, somewhere was – is – getting a raw deal. As a journalist, I was never invited to visit factories that produced collections for fast fashion brands. But I have visited a number of others. In 2009, I spent an afternoon at French brand Aigle's factory in Châtellerault, watching skilled craftspeople take turns in the 60-step process of making their natural rubber boots, by hand. I've had mine for 12 years – and have not bought a new pair since.

Four years later, and some 5,000 miles away from rural France, the Rana Plaza factory in Bangladesh, which produced clothing for many global fashion retailers, collapsed, killing 1,134 people. It was a wake-up call, shining a light on the appalling and dangerous working conditions of the people who make our clothes. The people we never see, the people we never think about. Lack of transparency in fashion supply chains is not limited to fast fashion brands; by no means does expensive equate to saintly status.

The same applies to fashion's environmental impact. Total greenhouse gas emissions from textiles production stands at 1.2 billion tonnes each year.[5] It relies predominantly on non-renewable resources, including oil to produce synthetic fibres, fertilizers to grow cotton and chemicals to dye garments. And it uses around 93 billion cubic metres of water annually. Speaking of which, when we wash our clothes in a machine, half a million tonnes of plastic microfibres are shed from synthetic fibres such as nylon, polyester and acrylic every year – a major contributor to ocean pollution. In fact, 35 per cent of primary microplastics entering the ocean are released through the washing of textiles.[6] If we don't change the way we choose to live, there will be more plastic in our oceans than fish by 2050.

The way I shop and the clothes I choose to wear have changed dramatically. In the last two years alone, I bought one new item of clothing (and it was made from recycled materials). I'm not going to write a list of good and bad brands for three reasons. One, there is no such thing. Even Yvon Chouinard, founder of Patagonia, says so, preferring the term "responsible" to "sustainable", but there are many well-known brands and retailers from which I would no longer ever purchase. Two, that would suggest our work is done: it's not. Three, I'd be taking away the thrill of the hunt; of discovering new brands, new businesses transforming the way we dress, new technology and new ways to challenge the meaning of "new".

Instead, I'll say this. Next time you decide to buy a pair of jeans, check you don't already own a similar style – and, if you do, fall in love with it all over again by shopping from your own wardrobe first. If it's a standout, seasonal silhouette you're after, ask yourself if you'll really wear it beyond next season. And if not, try renting it instead. Whether it's to satisfy a need or a whim, do your research before you part with money.

And read the following essays in this chapter. I first met one of the authors, the designer Christopher Raeburn, ten years ago when I interviewed him for *Drapers* magazine. He is as revolutionary now as he was then, an intelligent designer whose technical innovations in reappropriated and recycled fabrics results in clothes everyone wants to wear. Tamsin Blanchard, a fellow journalist (and many other things), is a leading voice in responsible fashion, reporting on and shaping the industry we know today. Meanwhile, Eshita Kabra-Davies, a self-confessed lover of fast fashion, decided to do more than simply stop buying countless items of landfill-destined clothes: she launched a fashion rental business instead.

Few of us buy clothes because we need to. Fashion is nothing if not fun. The thrill of a new purchase, season-defining silhouette and dress that garners a hundred compliments – my gran felt them all, without accumulating all those jeans. We do have to give up some things. We have to buy fewer clothes. We have to question where we buy them from. We have to embrace new ways of expressing ourselves through fashion.

But we don't have to give up on the thrill.

7

Eshita Kabra-Davies

"If Not Me, Who? If Not Now, When?"

Every summer until I was the age of 16, our parents sent my siblings and me to our hometown in Rajasthan, India. We had moved to Singapore when I was two years old and there was a worry we wouldn't understand our mother tongue, Hindi, or, worse, our heritage and culture.

I loved summers in the desert. A chaotic reunion for all 11 cousins under our maternal grandmother's roof. On some lucky occasions, I was allowed to shop at the local fashion market, which consisted of family-owned stores passed down from one generation to the next. The family at the store always sat you down with a cup of *chai*, asked you about your needs and then presented options while explaining the origins and artistry of each item. No, this was not an upmarket store or special purchase, nor was it an upmarket town. It's just how the local communities approach fashion and artisanship; they respect each other and the work that goes into one's livelihood.

The majority of my shopping experiences have been very different. From countless malls on Orchard Road in Singapore to the free delivery and discount codes at various e-tailers in the UK – there was no further sentiment beyond "fear of missing out". Often not able to logically explain why I *needed* new clothing, I also had no idea about their origins (besides the "made in" labels), nor did I reflect on how these items could be so cheap. And so my love for fashion turned me into a shopaholic – less glamorously known as a consumerist.

In February 2019, after more than a decade, I travelled to my motherland for my honeymoon as I wanted to reconnect and share my roots with my English partner. While I have many beautiful memories from our trip, I also remember the poignancy I felt when I realized a fundamental issue with one of my greatest passions: fashion.

Beyond the documented overflowing landfills of India, animals that I consider sacred as a Hindu were feeding on synthetic textile castaways in my suburban hometown. I felt guilty: I too had indulged far too many times in clothes I did not wear enough. As a proud Indian, I felt hypocritical: defending my people and demanding more respect for them, but also taking their labour for granted and polluting our country with my has-beens. I vowed thereafter to change my consumption habits and bring the concepts of community and sharing to the average consumer.

Two months later, alongside my full-time career in investment management and with no experience, contacts or influence in fashion, I founded By Rotation, the UK's first peer-to-peer fashion rental app. The concept is simple: to enable a diverse community to rent and lend designer pieces to and from each other. You can monetize investment pieces and access designer items at price points that compete with fast fashion. The end result is a self-sustaining and conscious community allowing you to enjoy the thrill of "new", without actually buying new.

The sharing economy has taken off in many forms – from Airbnb to Uber – so it makes sense for fashion to be next, with concepts such as resale projected to be worth $36bn globally by 2025.[7]

Fashion rental is not a perfect solution: inventory, dry-cleaning and transportation all have a detrimental impact on our planet. These aren't factors we deal with directly: By Rotation doesn't own nor hold the product – it acts as a marketplace and recommends eco-friendly cleaning and delivery solutions instead. As various designers experiment with biodegradable fabrics and new technology, I strongly believe that we also need to shift our mindset around consumption. Put simply, there are enough items of clothing in the world – far more than we need or wear – for new pieces to be made from scratch.

What is the purpose of fashion after all? It is a form of self-expression. It can make us feel empowered, more in touch with our inner self. But we have become disconnected from the people who make our clothes and the impact our consumption has on our fragile planet. It is vital, therefore, to bring people together and make them feel included – and fashion is my tool. No one can do everything, but everyone can do something. So, consider every fashion purchase beyond its price tag: it's in your power to vote with your wallet (without spending more) and support a more sustainable future.

8

Tamsin Blanchard

Lessons from My Mother

My mother was busy, sociable; a restless spirit with a strong sense of social justice and ethics. She was also infinitely stylish. She loved her clothes – they spanned four wardrobes, two chests of drawers and spilled out onto a glossy green bentwood chair. She had boxes and boxes of shoes and a selection of handbags, mostly black, leather and classic in style.

Her wardrobe took her through several decades, two children, two husbands, two grandchildren and a career that went from shipping clerk and part-time model to mature art student, lecturer, illustrator, children's author and Chair of the local Liverpool Community Health Council.

Arlene Blanchard – Min, as we called her – made a lot of her own clothes. She would use her trusted Bernina sewing machine to whizz up a shirt dress, or something quite ambitious like a voluminous asymmetric coat made to a Kenzo pattern from *Vogue*.

Min taught me to sew. She helped me to make an electric blue wool trapeze coat that I wore to school in my teens. Lined in shiny satin with a shawl collar, it stood out from everything everyone else was wearing and looked rather showy for my secondary school.

My mother had a knack of dressing in a way that was truly timeless. Although many of her clothes were 10 or 20 years old, they never looked dated. She favoured black and navy. She liked skirts to sit either just above the knee or below the calf. Her trousers were narrow, her jackets slimline. The silhouette was sleek, never fussy. For a day at the drawing board, painting, she would wear comfy denim dungarees, which I now wear. If she was teaching (on an Art Foundation course), she would always look practical yet chic; a tunic or smock over a pair of leggings. For evenings, something a little more glamorous, an asymmetric dress, perhaps. An insect brooch picked up at a flea market.

Many of her clothes were high street purchases – Chelsea Girl from the 1980s, St Michael knitwear from the 1970s. And there were also the pieces she made, as well as some more expensive ones I gave her, including a classic black Helmut Lang dress. She would have happily bought from Armani, Yohji Yamamoto and Jil Sander if she had the money. But she lived within her means, yet looked as though she wore those designers. She dressed up items collected from vintage stores – kimonos, 1920s dresses, even the odd Victorian lace jacket. I've taken to wearing a few of her things, such as the Kenzo coat she made, though sadly the Lang dress is too small.

My mother rarely threw anything out. When I came to clear her wardrobes a few years ago, after she passed away, it was hard to know how old anything was. It wasn't just that she had a style and stuck to it – she looked after her clothes so well. They were dry cleaned and kept pristine in their polythene bags. She taught me to dab a little clear nail varnish on the back of a button to stop the threads coming loose (and to stop a hole in a pair of tights running).

Every single piece in her wardrobe was loved – but it was a lifetime's work. My mother didn't consider herself to be part of a sustainable fashion movement, because she always understood that her clothes were precious. She was born during the war, when nothing was wasted and everything had a value. The clothes she made as a young woman ensured she could feel the quality of fabric on a roll. Above all, she had respect for the craft of making clothes. She knew how long it took to create a dress or a jacket. The effort she put into making her clothes and looking after them, repaid her with a lifetime of looking effortlessly cool and stylish.

Three things I have learned to pass on:

1. Find your style: understand what colours and proportions suit you. This means you will never buy something on a whim. Min never bought anything she didn't wear.
2. Learn to sew or find a local tailor. When clothes fit properly they always look good. You will never again take for granted the people who make them.
3. Respect your wardrobe. Min's shoes were kept neatly in boxes, jackets hung next to each other, long dresses on a high rail. Your wardrobe then functions for you: you can find things easily and won't buy more of the types of clothes you already have.

9

Christopher Raeburn

Make Good Use of Bad Rubbish

It was the Wombles who advised us to make good use of the "things that the everyday folk leave behind". Wombles are not so well known outside the UK but I think they should be. Looking back, I owe much of my career to these fictitious, pointy-nosed, furry and rather inventive characters created by Elisabeth Beresford.

For those not so Wombles-savvy, look them up online or at their home of Wimbledon Common in London. They had a mantra – "Make Good Use of Bad Rubbish" – and each TV episode focused around the Wombles scouring the common for litter and discarded objects. The objects were then repaired, recycled or creatively transformed for another use – a notion that resonates as powerfully today as it did three decades ago to my curious younger self.

RÆPAIR. CAN I FIX IT?

More than 100 billion pieces of clothing are made each year worldwide; an unimaginable number that stands testament to the forces of a capitalist society. Similarly astounding yet more conceivable is that one lorry full of textiles ends up in landfill every second[8] – design for longevity is no longer commonplace in a world of fast-moving trends and testing bottom lines.

At RÆBURN we believe that a tactile approach such as repairing fosters a stronger connection to and appreciation for material value – qualities that are distinctly missing in a fast fashion world. Not only will a repair extend the physical life of your item, it will add an emotional durability to the product, elevating its story and personal sense of worth. All is in the power of a needle and thread; two simple and accessible tools, with a pinch of enthusiastic determination. Further still, consider when a repair becomes a product enhancement – then we've got a very exciting concept. Upcycling and rejuvenating clothes instead of discarding them is a significant way for anybody to reduce their impact and be more responsible.

RÆMADE. NEW, BUT OLD?

RÆMADE was born during my graduate collection at the Royal College of Art. The geek in me was fascinated by the fact that if you wanted to buy the original materials on a roll, they were either scarce or expensive, and yet I was able to find original, unworn pieces in places like Portobello Road antique market. Close to 15 years since my leaving the Royal College, we have built a business using predominantly existing materials like parachutes, life rafts, fire jacket, diving suits and 1950s silk maps; there's very little we can't take and make into something contemporary, wearable and above all useful again.

The very nature of the fashion industry tends to promote an insatiable appetite for new; a scary but enticing formula that helps to drive our consumer culture. If we rethink the model and have the courage to assign value to obsolescence, then we have an opportunity to make "new" from old. A designer's obligation is to scrutinize what they make and how, before pen meets paper. As consumers, there's an opportunity to support those who work responsibly, but, culturally, we have a bigger obligation to consider: do we need anything new at all?

RÆFOUND. WHAT COULD BE MORE RADICAL THAN MAKING NOTHING AT ALL?

Amid lorry-loads of textile waste, we now face a post-Covid-19 era with a new environmental and inventory crisis. From one pandemic to the next, firefighting existing issues while making more of the same won't help. We've built our own prison in which there's a race to the bottom where overproduction and consumption are symbiotic – the repercussions for our people and our planet are real, visible and devastating.

So what could be more radical than making nothing at all? Rather than designing new among a flood of old clothes, can we repurpose what already exists? Yes. As an evolving non-seasonal range of personally sourced, original, unworn military apparel and accessories, RÆFOUND seeks to encourage us all to repurpose and reuse through any method. Look at marketplace platform Depop, whose community is positively producing, upcycling and selling more than ever. It is much more rewarding (and affordable) to do so.

RÆBURN has always sought the path less travelled. It's been a challenging journey, but one that has been the result of an optimistic community that shares our vision, and has supported us in our constantly evolving brief to push the industry forward. My positive instructions to any reader are:

1. Can I fix it?
2. Does it need to be new?
3. Does it already exist?

After all, I suspect if we were more like the Wombles, the world would be a much better place.

Three Things to Do

My best friend got married last year and she asked me to do a reading at the wedding. Even if she hadn't, I would still have cared about what I wore. It was autumn, and the most weather-appropriate dresses in my wardrobe were dark. It wasn't the look I wanted.

A few years ago, I may well have treated myself to a new purchase. This time, I shopped elsewhere: my sister's wardrobe. It was like being in a boutique, one where every dress was available in my size and edited to styles I like, but without the price tag. A boutique that didn't exist outside my sister's bedroom. I'd thought that limiting my purchasing behaviour when it came to clothes – having been a full-time fashion journalist – would be hard. Not at all.

Traditional retail has been replaced with an abundance of alternatives, many of which are much kinder to the people and resources behind the production of our clothes – and much more fun.

ACTION 1: PRE-LOVED FASHION

To do: satisfy the thrill of new clothes by borrowing, renting or buying second-hand instead.

- Why do you buy clothes? Because you love fashion? Because you want to try out new styles? Because it makes you happy?
- How often do you buy new clothes?
- How often do you borrow, rent and thrift shop?

ACTION 2: YOUR WARDROBE

To do: start shopping from your wardrobe first. You never know, you might find forgotten favourites or reinvent new looks.

- Could you name all the clothing you own? If not, organize your wardrobe to remind yourself – and create different outfit configurations.
- Are there any items you no longer like? Could you upcycle them into styles you do?
- Can you sew? If not, learn or find a local tailor. You'll save money by fixing existing clothes and learn to really appreciate the people who make them.

ACTION 3: CONSIDERED PURCHASES

To do: if you do buy new, do your research and support brands that care about the people and resources behind the making of your clothes.

- How much do you know about the working conditions of the people who make your clothes?
- How much do you know about the materials from which are they made?
- How much do you know about the resources used to make them?

TRAVEL

- Outdoor air pollution kills an estimated 4.2 million people each year, with 9 in 10 of us living in places where air quality exceeds the safe guideline limits from the World Health Organization.[1]
- In terms of an individual's annual carbon footprint, a single long-haul return flight can generate more emissions than any other activity in a given year.[2]
- We can't see these particles, but exposure to fine particulate matter of 2.5 microns or less in diameter (PM2.5) can cause cardiovascular and respiratory disease, and cancers. Tailpipe emissions from road transport account for up to 30 per cent of fine PM in urban areas.[3]

"Most human decisions are based on emotional reactions", writes Yuval Noah Harari in *21 Lessons For the 21st Century,* a study of the issues facing the future of humanity. Similarly, when speaking about our relationship with climate change, the Nobel Prize-winning scientist Daniel Kahneman said: "To mobilize people, this has to become an emotional issue."[4]

To act, we have to *feel*.

I used to live two miles from work. I would cycle there, along the river, knowing I had an enviable commute by London standards. One day, I decided to drive because it was raining. So did everybody else. We didn't want to get wet. We didn't want to navigate the awkward public transport that required at least two buses. A journey that normally takes less than 15 minutes by bike took 45 minutes by car. I sat in traffic for most of it, watching parents walk their children to school along a road that regularly breaches safe air quality limits. The children carried umbrellas.

Almost all of us – 99 per cent of the global population – are exposed to air pollution levels that put us at increased risk of heart disease, stroke, chronic obstructive pulmonary disease, cancer and pneumonia. A leading cause is particulate matter (PM) emissions – particles that are too small to see but can settle anywhere in our bodies, including the brain.[5] PM mostly originates from engines.

I know this because, when she was four years old, my daughter was diagnosed with asthma. Soon after, I joined a local clean air campaign group called Mums for Lungs. I vowed never to sit in unnecessary traffic again. I'd carry an umbrella instead.

During that time, I volunteered on behalf of Mums for Lungs to write a letter to Heathrow's CEO and the relevant MPs to oppose the airport's proposed plans for a third runway. Partly, I did it because I live under the Heathrow flight path. It's annoying to be woken up at 5am. But it's nonsensical to propose a third runway during a climate crisis. I read the consultation, intended for the public but filled with jargon and nonsense, including the claim that Heathrow's owners are "committed to developing Heathrow responsibly and sustainably, with a focus on the wellbeing of our communities and the environment. We will only grow within strict environmental limits." Yet environmental limits were already exceeded even without expanding Heathrow and increasing flight numbers. Heathrow is the largest single source of carbon emissions in the UK.[6]

I won't bore you with further claims. In fact, amid all those pages, nothing hit home as much as this: the new third runway would add an extra 700 flights per day. Have you ever wanted to fly somewhere and not been able to? Have you ever needed a choice that would justify 700 additional flights per day? Me neither. As with many decisions that affect our planet, they benefit the few at the expense of the many.

In February 2020, campaigners won a landmark Court of Appeal ruling over Heathrow's plans on environmental grounds. Then in December of the same year, the Supreme Court overturned the February judgement. At the time of writing, the work is not yet done, but each collective act – be it a letter or the lawyers who initially brought the case against Heathrow – is buying time.

Everyone can make a difference.

10

Rosamund Kissi-Debrah

A Breath of Fresh Air

The second after a baby is born, there's an anxious moment: everyone involved in the delivery holds their breath until the baby lets out its first cry. The most crucial sign that all is well: the baby is breathing. Yet after this pivotal juncture, we rarely give breathing another thought. Certainly I didn't, especially when I gave birth to Ella Roberta, a healthy bouncing baby. We take breathing for granted. But since my daughter's passing nine years ago, from a rare and severe form of asthma linked to air pollution, I've learned the importance of not just breathing, but breathing clean air.

Humans can survive three weeks without food, three days without water, but we will only survive three minutes without breathing. In international law, our human rights to access and drink clean water are protected. Our rights to food and shelter are protected. But not our right to breathe clean air. According to the World Health Organization (WHO), every year seven million people die prematurely worldwide due to air pollution. Dirty air is linked to asthma, lung disease, diabetes, cardiovascular disease, dementia, cancer and many more illnesses. Every ten seconds, someone has a potentially life-threatening asthma attack, in the UK alone. The air we breathe impacts our lives from cradle to grave.

Ella was an incredibly bright and healthy child – she was always swimming and won gymnastics medals – until she developed severe asthma that took her to hospital 28 times in three years. For seven years, I fought for an inquest into Ella's death. Finally, in 2020, a coroner made legal history by ruling that air pollution from traffic near our home – in Lewisham, London – contributed to her death.

The UK has been in breach of legal limits of air pollution since 2010, the year Ella first became seriously ill with life-threatening asthma. Experts estimate that air pollution in the UK is responsible for 36,000 premature deaths every year, 9,000 of which are here in London, where my family and I live. Yet if the UK government adopted the limits set by the WHO, it could save up to 17,000 lives every year. So why doesn't it? Because the fossil fuel industry has enormous power and influence over governments worldwide.

The ambition from the WHO is to reduce the number of deaths from air pollution by two thirds by 2030. But, in my opinion, that's not ambitious enough. Who is the "one third" excluded from the target?

As we gain a better understanding of how detrimental air pollution is to our health, the more pressure we can – and must – put on our government. This needs to be a joint effort, and there is much that individuals can do. Supporting organizations that can lobby government is one option. Client Earth, for example, has taken the government to court three times over its failure and inaction to clean up our air and has been successful each time. We need a transport infrastructure in place that ensures our buses and taxis are all electric, and that cycling is safe and prioritized. New diesel cars must be outlawed by 2025. And everything begins with good education – much of the work we do at the Ella Roberta Family Foundation supports this.

During lockdown, we saw how quickly our air was cleaned up. In London, levels of the pollutant nitrogen dioxide (NO_2) were reduced by 55 per cent due to less road traffic, according to research from King's College, London.[7] Of course, we were forced into it, but it shows the immense power we have – when we use it. Government data shows that for distances of one to two miles over 60 per cent of journeys are made by car, rising further for journeys under five miles[8] – journeys that could easily be done by foot, bike or public transport. So here is something that those of us able to do, could do right now: every time we make a journey under five miles, let's do so without a car.

Ella is now (at the time of writing) the only person in the world to have air pollution listed as a cause of death on her death certificate. Following his landmark decision, the coroner made it clear that air pollution is a health crisis and governments need to approach it as they would any other pandemic that is killing people.

We can see when water is dirty. We can see when food is rotten. But we can't see – not in the everyday – when air is polluted. Breathing is innate. We may not be able to see it, but taking a breath is the most precious action we take.

11

Juliet Kinsman

On How to Travel

Travel has always been my way of being. We moved to Algeria from Canada when I was six weeks old. Then New York, because of a diplomatic dad, followed by Washington DC. Teen years were in Bucks and Somerset, followed by extended spells in India, Greece and Bali. I've been clocking up air miles my whole life – a peripatetic lifestyle extended from childhood into a career as a travel writer. Across five decades, I've seen globetrotting go from glorified to vilified, to quite simply not being an option. It's an undisputed fact that by setting foot on a plane, we are boosting our personal carbon footprint through supporting a form of transport that burns fossil fuels by the tonne. But I don't believe banning international holidays is an effective solution to the world's climate problems; the key to sustainable travel is to travel better.

What a journey it's been for humanity since the term "jet-set" was first used by journalist Cholly Knickerbocker in the 1950s – the pen name adopted by Igor Cassini for his gossip columns. Decades past, it was only the super-rich who hopped from New York to London by air. The only polar bears or melting ice likely to cross their minds was when they hoped turbulence wouldn't spill their on-board scotches-on-the-rocks on their white fur coats. Scroll forward, and a mention of bears or glaciers will likely stoke guilt rather than glamour when we picture Arctic animals losing their habitats as a direct result of the world heating up.

We constantly hear talk about the climate emergency – yet still not everyone gets it. Carbon is the real star – well, antihero – of this epic. It plays a lead part in the globe-swaddling blanket caused by the layer of greenhouse gases in the atmosphere. Constantly shape-shifting on and around our planet, and as ubiquitous an element as can be, carbon molecules display a wider repertoire than Christian Bale. Most living things are a composition of them – people, plants, animals – and the decomposition of organic matter over millennia creates fossil fuels. By burning them to propel our planes, we release CO_2 back into the atmosphere. Big time. The Committee on Climate Change predicts the aviation sector to be the single biggest source of emissions in the UK by 2050.[9] Your seat on an aircraft acts as the single biggest contributor to your footprint.

What's more, those worst affected by the climate emergency aren't even flying; estimates suggest only between 5 per cent and 20 per cent of the global population have ever been on a plane. So we owe it to the remaining 80 per cent to 95 per cent to mend our ways.[10]

The pandemic propelled us to change the way we think about travel. We've slowed down and I would hope we've emerged from the hiatus in air travel more conscious, conscientious and sensitive to the health of people and planet. And, an optimist would assume, eager to help the world heal through our holidays.

We will travel less, but we will travel better. *Flygskam* is the name the Swedes gave the flight-shaming movement in 2018. The antidote is *tågskryt* – train-bragging. Now that flying isn't as easy and cheap as it used to be, let's follow road, rail or water routes – and enjoy the journey becoming part of the holiday.

Of course we'll still fly, so look for airlines that invest in renewable fuels. Scandinavian airlines tend to be more sensitive to aviation-improving innovations. Book routes with newer aircrafts for more fuel-efficient transits. "Net-zero" is when a business measures emissions from all their activity and balances it through offsetting: for every tonne of carbon emitted, they avoid, sequester or capture the equivalent in another location. Many airlines are striving for "carbon neutral" status by offsetting all operations and halving emissions by 2025. Clearly, it's cannier to slash emissions in the first place rather than to pay a do-gooder to compensate for wrongdoings. So try to cut out the middleman.

I'm not saying I don't instantly go weak at the knees thinking about sipping rosé in Provence or picking at tapas in Barcelona, but those weekends away on a whim that rely on short-haul flights are the killer – it's the taking off and landing part of flying that's most polluting. Hopping on those short-leg flights so regularly had become too habitual. Can't resist that cheap flight to see family or for a special occasion? At least factor in an overland or across-water journey for one leg.

The truth is, it goes deeper than that. International travel is vital to many under-privileged communities in a world where one in ten people are employed in tourism (according to the WTTC).[11] Entire communities in Asia, Africa and the Americas rely on our holiday spending as a form of global wealth distribution, which unlocks education and healthcare and improves infrastructure. The rainforest-rich island of Madagascar, for example, has one of the highest levels of biodiversity in the world, but its species are under severe threat from the effects of the climate collapse and deforestation. As a nation that contributes very little to the climate crisis they are also experiencing the world's first climate-induced famine. How can you help? By booking a holiday there with a sustainable hotel.

I urge you to think beyond a bargain-basement package holiday to a foreign-owned hotel chain, which relies on cheap labour and imported ingredients. Consider, instead, community-based tourism that focus on experiences hosted by local residents in a way that does good for the whole area. The reward is a deeper, more meaningful and textured insight into local cultures. Sign up for excursions with social enterprises, or turn to ethical operators who have planned trips in this way to treat travellers to a truer taste of far-flung cultures, while benefiting those who live in more remote destinations.

Sure, this could mean radically reducing the number of holidays we take, but that in itself carries benefits beyond reducing our carbon footprint. Dr Jeroen Nawijn's study at the Centre for Sustainability, Tourism and Transport in Breda University in the Netherlands declares the anticipation of our holidays to be the part that gives us the most joy, releasing more uplifting endorphins than actually being away.[12] So take a deep breath, savour the slowdown, and salute long-haul forays being something less frequent, but special again.

12

Jona Christians

Motoring toward New Ideas

We were in the last year of school – my friend Laurin Hahn and I – and the question of "what next?" was looming. One day, when we were discussing it on the phone, the conversation turned to climate change and, specifically, to our dependency as a society on crude oil – the world's largest energy source. Could we play a part in finding solutions to that? As we started to research different industries, we learned that oil is the dominant source of energy for the transport sector. But we knew nothing about cars; we didn't even own a car.

Back then, in 2012, electric vehicles were a rarity. And if you happened to see one, it was a rather funny-looking small car with a short range. There were rumours about a

Munich car company bringing out its first electric vehicle and another, in the US, was selling electric sport cars. But nothing was readily available in the mainstream.

We had little to go on, but if we wanted to be part of the solution, creating a solar and electric car was the logical option. So we began working on a project to build a prototype in secret. We didn't tell anyone. We'd sneak out of class to call suppliers at 10am on a weekday so that they'd think we were professionals, not school kids.

A key finding in our research was that the average car sits parked on the street or driveway 95 per cent of the time.[13] That posed two important questions: first, if we consume resources to make a new car, how can we at least make sure that these resources are not being used for just 5 per cent of the time? Second, why have so many cars in the first place, when they're barely being used?

Our initial idea quickly developed beyond the car as a product; we wanted to reimagine the concept of what a car could do. What if, instead of private ownership, people shared their cars within their communities? What if, when fully charged and stationary, a car became a hub for storage and distribution of electricity to other vehicles, to your home, to the grid?

That ambition – to create a car that could change a broken system; that could change mindsets around status and ownership – led to the three fundamental principles of Sono Motors: the integration of solar power into the whole body of the car in order to make it less dependent on the charging infrastructure; a bi-directional charging system to store and give back energy; and integrated shared access built into the design of the car.

After three years working in the garage, we created the first prototype – the Sion – and launched our crowdfunding campaign. Up until the moment when we pressed the button to go live, we had no idea if anyone would respond. But we surpassed our target by 125 per cent with 2,000 supporters, including 16 people who even paid the full price of the car upfront. And what was important for us at that time was to be in really close contact with these people. So we wrote and talked to many of them. We listened to their feedback and we included them in major decisions about the design of the car, charging speeds and a lot more.

The next step was to build a team of motivated and skilled people. Again, knowing nothing about the structures of traditional companies, we did what felt right in the moment. We had team meetings in which every single person participated once a week; we openly celebrated failures so that the whole company learned from all our mistakes; we had a deep trust in each person's core motivation. To begin with, most of us were in similar positions – young, independent, no ties. But I remember the moment when the first person with a child joined Sono. It felt different. There was a family behind this person. It gave us a strong sense of responsibility.

Today, we can offer our solution for affordable and practical mobility. The Sion is a five-seater family car with a spacious trunk and tow bar. Integrated into the body of the car are 248 solar cells that can add, depending on the weather, 112km of driving range on average per week to the car's 54kWh battery. This creates full self-sufficiency for short distances. To charge the Sion to 80 per cent takes around 35 minutes at a fast charging station, and it can also be recharged at any public charging station in Europe, via your regular home power socket or from another Sion. Its on-board bi-directional charger lets you share your Sion's power to recharge other electric vehicles or devices, while our bi-directional wallbox enables the

Sion to become a sustainable power plant for the home. We've received 16,000 reservations so far, and the Sion is due to be delivered in 2023.

Our goal is to have every car on the street electric and shared. Luckily, we're part of an industry that is relishing this challenge and accelerating technological innovation to make electric vehicles (EVs) the mainstream – and helping decarbonize our planet. Some people could make the switch to an electric car today. For many, that's not yet financially viable, but as an industry, we are working hard to make EVs more accessible to all. In the meantime, leasing options could be a solution. If you don't own a car, could you share your neighbour's? Or use a car-sharing scheme? Or consider short-term and tailored rental?

Looking back at the beginning of our Sion journey, looking at the whole picture, I believe that there was something bigger taking place – both back then and at this very moment. Bigger than the story of individuals. Bigger than the story of Sono Motors. I believe that what happened at Sono was that a group of people tuned into something that was in the air already. There was no magic, no special people – just individuals who did what felt right. Because the thought of not doing it simply wasn't an option.

Three Things to Do

The distance between my house and the local swimming pool, where my daughter has weekly lessons, is 1½ miles along a single road.

Even though parts of that road are lined with cafes and independent shops, it is used as a thoroughfare by cars and trucks; it's a polluted stretch that often resembles a car park. I'm grateful for the cycle lane, but, between the traffic and pollution, my daughter is too young to cycle there. So I mapped out an alternative cycle route, through the common and quiet residential streets.

Some days, it doesn't take much longer, given that cars and buses often sit in traffic. My daughter complains sometimes, but I don't budge (unless it's pouring with rain). I don't do it just to avoid the traffic and pollution. I do it so that I try to lead by example.

ACTION 1: ACTIVE TRAVEL

To do: if you are physically able, for journeys under five miles walk, cycle or take public transport.

- Next time you're about to get in the car, if it's a short journey, could you take a different and cleaner method of transport?
- If cycling is an option, try cycling apps that plot bike-friendly journeys, away from busy roads.
- Support organizations that lobby government to clean up our air and create appropriate transport infrastructure. You could find a local campaign group or lend support to larger organizations such as Greenpeace, Friends of the Earth and Client Earth.

ACTION 2: AIR TRAVEL

To do: fly less, and travel better. Do the research into destinations, airlines and accommodation providers before booking your next holiday.

- When booking a flight, check that the airline invests in renewable fuels. Do they use newer, more fuel-efficient aircrafts? Are they striving to become carbon negative or climate positive? These go much further than "carbon neutral". Beware of greenwashing.

- Could part of your journey be done by train or boat, rather than by plane?
- For your next holiday destination, could you book it through an ethical operator that supports community-based tourism and social enterprises, for a truer taste of the local culture?

ACTION 3: CAR TRAVEL

To do: if you can, choose alternatives to private cars that run on petrol and diesel.

- For private car ownership, look into electric models and leasing options (the latter offer flexibility and cost-savings).
- How often do you use your car? Could you share it with a neighbour or use car-sharing schemes?
- Are you considering buying a car, either for the first time or to replace your existing one? The pace of innovation in design and technology will make electric vehicles more reliable and affordable in the near future – could you wait?

SPEND

- The world's 60 biggest banks have provided $3.8tn of financing for fossil fuel companies since the Paris Agreement in 2015.[1]
- Eighty-two per cent of the emissions from a smartphone come from its production,while electronic waste has become the world's fastest growing waste stream.[2,3]
- $47 trillion is invested in pension funds globally.[4] In the UK, modelling by Friends of the Earth suggests that £128 billion could be invested in fossil fuels across pension funds and schemes, an investment of £1,916 in fossil fuels for every person in the UK.[5]

Catherine Chong suggested we meet for coffee at a café in London's Borough Market. As she approached me, her face lit up like the golden bike helmet she was carrying. I was standing by the bakery's window display, crammed with perfectly formed pastries. Catherine hugged me warmly – this was the first time we'd met – threw her arms around the waitress and ushered us both to a tiny outdoor table, facing the pastry-filled window. "What do you recommend?" Catherine asked, clasping her hands together, eyes wide with delight and anticipation. The two women embarked on a detailed discussion of flaky pastry layers, sweet-versus-savoury fillings and a warm, nutty-brown butter cake. I ordered the brown butter cake.

"Here, you have to try some of mine," Catherine said later, slicing her pastry in half and pushing the plate toward me. I looked at my own plate, and saw cake crumbs.

I wasn't expecting all this joy, this love of food, this sharing of plates. I was expecting to meet an investment expert, someone who could talk to me about pensions. I'd brought with me all those clichéd preconceptions.

Catherine is – among many other things – an impact and ESG (ecology, social and governance) advisor; a consultant for global FMCG groups, private equity and sovereign wealth funds. She advises businesses on how to support positive ecological and social impacts. Projects in her portfolio include rainforests and wildlife conservation, sustainable crops certification, ethical lending and investing, and sustainable product strategy in consumer goods.

But her passion is food, partly because she witnessed first-hand the devastation that poor investment can wreak on farming, when her father lost his agroecological livelihood to industrial practices. It's a heart-rending story. She grabbed the opportunity to study law after winning a scholarship and, later, studied climate and socioecological economics to work in the financing of sustainable development.

What's all this got to do with pensions? A report by Feedback found that global meat and dairy companies received over $478 billion in backing from investment firms, banks and pension funds between 2015 and 2020.[6] The report says that "there is no version of industrial animal agriculture that is compatible with climate justice and a zero-carbon future."

But Catherine's ambition is for us to look at impact investment as a way of life. Whenever we spend money, however little the amount, and on whatever product or experience, we are directing a flow of capital to particular production systems and supporting them – even if subconsciously. It could be the chicken you buy to cook for dinner, the restaurant you dine in, the pension fund you invest in.

It is all connected.

But it doesn't help that people like me view financial investments in the abstract, on a spectrum that begins with boredom and concludes with complicated. Like Alasdair Roxburgh, Chair of Friends of the Earth Europe, I chose my bank when I was 18 based on the fact that it would give me a free railcard in return. Aside from its high street name, I knew nothing else about it and made no further enquiries before letting it manage my money for the next 20 years.

A report from 2021 by a coalition of NGOs found that the world's 60 biggest banks provided $3.8tn of financing for fossil fuel companies since the Paris climate deal in 2015. Globally, US bank JP Morgan Chase was named the worst offender. In Europe, Barclays was the biggest fossil fuel investor, while French bank BNP Paribas was the world's single biggest supporter of offshore oil and gas.[7]

Just imagine if that colossal amount of money was invested elsewhere.

One obstacle is trust in the familiar. The money we work hard to make is precious. In the 20 years I'd banked with Barclays, nothing went wrong. And I hadn't even heard of Triodos Bank (see page 114) until a few years ago, despite it being more than 40 years old.

We can be reluctant to deviate from the mainstream. There's comfort in the familiar. But there's reward in discomfort.

This is what Tessa Wernink discovered when she turned down a "sensible" job offer to become the co-founder of Fairphone. Smartphones are problematic for two reasons: the conflict minerals mined for their manufacture and the subsequent waste they create. According to *Which?* magazine, most electronics contain more than 33 per cent of the 92 naturally occurring elements. Four of these – tungsten, tantalum, tin and gold – are known as conflict minerals, because their theft and sale have been linked with killings, violence, rape and other human rights abuses in the Democratic Republic of Congo (DRC) and other conflict zones.[8] Tessa wanted to do something about that.

Our instinct, particularly when it comes to investment, is to follow the path most travelled. To bank with high street names, to buy personal tech from the tech giants, to let well-known pension funds choose our portfolios.

The experts in this section are united in their advice: explore the alternative. By doing so collectively, we pave the way for the paths less travelled to become the paths most trodden – and change the system as a result.

13

Alasdair Roxburgh

Money Matters

To one side of a small courtroom in Belfast, Fidelma O'Kane sits alongside her husband and a huge stack of papers containing months of research and data. Fidelma is representing the local community of County Tyrone in Northern Ireland by challenging the decision to extract gold from the Sperrin Mountains, on the grounds that mining would damage the environment – an Area of Outstanding Natural Beauty. Fidelma has never before been involved in activism; she is a grandmother who simply wants to protect her country.

On the other side of the courtroom sits a dozen people: lawyers, representatives from the council, from the mining company. A real-life David and Goliath. "I've read this book," Fidelma tells the judge, holding up a copy of *Judicial Review in Northern Ireland*. The judge stops her. "Sorry to interrupt," he says, pointing to a man among the dozen representatives, opposite Fidelma. "I'm not sure if you're aware, but the man who wrote that book is in this court today, representing the people you are challenging."

Never has blood drained from an individual's face as fast as it does from that man's. And Fidelma, quick as a flash, like a seasoned barrister, turns to him and says: "Thank you very much for writing it. It's been so helpful. Maybe you can sign my copy at the end?" Hysterical laughter erupts in the courtroom. Was that a smirk from the judge, too?

Fidelma was hoping to get a judicial review on a couple of grounds; she won that judicial review in 2017 on seven grounds. This is an individual who wanted to stand up for her community. Not a lawyer, with years of experience. That story, that image, that exchange will stay with me for the rest of my life. That's what gets me out of bed in the mornings.

We often question the power that individuals have to effect change. But if there is one area where people can make a difference, it's money. If we change how we use our money, we create systemic change.

How did you decide who to bank with? I chose my high street bank based on the free railcard they were offering me at the time, when I was 18. I stayed with them for years; nothing ever went wrong – that I could see, at least.

Our relationship with the financial industry often lacks depth, in part down to education (or lack of) in schools. Much like how sex education used to be taught – if you have unprotected sex, you'll either get pregnant or die, was the message, before we moved swiftly onto English – we learn very little about how money works. It's no wonder, then, that our childhood perception of a bank is often indistinguishable from a big vault. Money goes in, gets locked away and the same money comes back out again.

We move on from the vault image in adulthood, but how many of us have opened a bank account and looked at where that money will go? How many of us have asked a bank manager about the bank's approach to investment? About its business plans? It's not something we do. We're looking for somewhere that will keep our money safe and secure, with good access and perhaps additional support if we're struggling.

We've worked hard to earn it, after all. For those in fortunate positions, £20 a week – the boost in universal credit that was withdrawn in autumn 2021 – may not sound like much, but that drop puts people into fuel poverty; it forces them to make hard choices between heating their homes or putting food on the table. Our money is precious. It needs to be looked after. We think, rightly, about our immediate needs.

Some people have the privilege to save money, both for the short and long term, perhaps for the future of their children, for future generations. The truth is, historically and to this day, our money helps to fund the climate crisis, whether that be coal power plants in Southeast Asia or mining operations around the world. All those investments have a negative impact on our climate and our environments.

Which is at odds with why we save in the first place. If that money is being invested in activities that compromise the future of our planet and humanity, what does that really mean for us? The science tells us that, if we don't reach net-zero by 2050 – for which the work needs to start now – we will be feeling the catastrophic impacts of climate change. I want my money to be invested in activities that help us reach net-zero; that tackle – rather than fuel – the climate crisis. Because in not doing that, I'm compromising my own future.

Furthermore, given this global urgency to move away from fossil fuels, investing in such industries will be bad investments. We're shifting our energy production away from coal and oil to renewables; we're switching to electric cars; we'll move away from gas boilers next. There is logic to banking and investing with ethical organizations.

If you are in a position to save money, you do have a choice in how that money gets invested. Having that luxury, being in that privileged position and making a good choice benefits us. More importantly, it benefits those who don't have a choice, who are not able to make those savings yet, who will be impacted first and worst by the climate and ecological crises. Think about where your money goes. And, if you can, switch to an ethical bank. That can be a starting point.

Just look at our eating habits. We've reduced our daily meat consumption in the UK to the point that we are now eating *17 per cent less meat* than we did ten years ago. In finance, collective individual action can stack up to systemic change. Banks such as Triodos invest in environmental, cultural and social enterprises that change the system. They're transparent in what they do. You can use their interactive map to see what they're investing in, right down to your local area. Think of the difference when trillions of pounds are invested in these projects, rather than in the fossil fuels industry.

The financial system rewards short-term rather than long-term growth. Why worry about what will happen in 30 years' time when you can make a lot of money in ten years' time. We are also talking about institutions here; these aren't agile start-ups. Change in institutions is chronically slow. Therefore, those who work in the finance industry have vested interests. That's why governments should be playing a role to help force change. And we need to keep up the pressure for them to do so.

We can start by asking questions of the financial institutions we choose to look after our money. What are my options?

Is a higher return worth it if it comes at the cost of the planet? That's a privileged question to be able to ask, and not everyone can. Search online, do some digging. Go to the consumer website Which?.com. Seek out advice, seek out support. Markets respond to consumer demand.

I grew up in a household with a strong sense of social justice. My mum took me to my first march when I was still in a pram. But the turning point came toward the end of the 1990s, when she took us down to Birmingham. "Everyone will be wearing red," came my mum's sales pitch, "so you can wear your Liverpool shirt." When we arrived, we joined tens of thousands of people to create a human chain around the city in protest. We were part of the Jubilee Debt Campaign in 1998 to eradicate Third World debt. In the 1970s, rich nations had invested money into all sorts of projects in developing countries. But when the oil crisis hit, all of a sudden they wanted their money back – and large interests on their loans. For the developing nations, it was crippling; they could no longer invest in education, in health. As a teenager in the 1990s, it seemed fundamentally wrong to me. That march was formative; it shaped my thinking, particularly around money and what it can do for good and for bad around the world.

You can drive yourself mad trying to save the world. You can shop less. Drive less. Eat less meat. But short of hiding your money under a mattress, you need a bank. And the financial system underpins all these decisions. If none of us took any individual action we would – and will – be in even more trouble than we are now.

14

Tessa Wernink

Explore the Alternative

After a few years of freelance work I'd been offered a job at a large corporate PR firm in Amsterdam. It was secure, sensible, stable. I was about to accept the offer when, quite out of the blue, I received a call from Bas van Abel with a very different proposition: to help him set up his business as a social enterprise. Bas worked for Waag Society, who I'd worked with on and off for a few years. I was fascinated by their approach to technology as an instrument for social change. One of their projects at the time was Fairphone.

The systems approach behind Fairphone is best explained by the 1980 speech of Nobel Prize winner Milton Friedman, who used the making of a pencil as an analogy for how our economy works. Do we know where the wood from which a pencil is made comes from? What do we know of the compressed graphite in the centre of a pencil? The rubber? The brass ferrule? Friedman said he hadn't "the slightest idea where it came from. Or the yellow paint! Or the paint that made the black lines. Or the glue that holds it together …"[9]

But while Friedman celebrated the invisible hand behind our economy, Fairphone questioned it. Few people knew how a smartphone was made. Once we began unravelling the process, we started to understand it. Once we understood it, we could make interventions. Waag Society had started Fairphone as a campaign to highlight both the use of conflict minerals in phones and, in turn, the extractive nature of our (capitalist) system. And now Bas had secured investment to develop Fairphone. That's when he called me.

I sought the advice of friends and family. "Take the PR job. You'll get good training; you can work for a start-up later," they told me. All, except my mum and partner.

It could have been a pencil. It could have been a chair. I accepted Bas's offer because, even before I asked other people what I should do, I already knew what I wanted: to swim against the tide. I wasn't sure if it was "wise", given my responsibility as a mother, but I felt excitement and determination to raise awareness around these issues by taking action. Showing there was a different side to things. After all, it was the American existential psychologist Rollo May who said that the opposite of courage in our society is not cowardice; it is conformity.

How your phone is made matters because the mining, assembly, use, disassembly and re-use of every material needed to create it has serious consequences for people and the planet, for human rights and natural resources. Our technology is designed from a user perspective. But materials don't stop existing when you stop using your phone. Fairphone wanted to extend our idea of technology beyond consumerism and start seeing the materials and people that make our products as valuable. So, rather than walk away from child labour and dangerous working conditions, from pollution and resource depletion, Fairphone set out to trace a complex supply chain and make it fair and transparent, aiming to use the phone as a platform to bring people together around shared values and changing the industry from within.

We tackled electronic waste – the world's fastest-growing domestic waste stream, driven by high consumption, products designed for obsolescence and lack of repair options – with design. When 82 per cent of a phone's emissions come from its production, making a phone that lasts makes sense. In essence, the fairest phone is the one you already own.

Yet phone providers, by offering subscription models that encourage us to upgrade our phones every two years, present the idea that the phone itself has no value; the value is tied up in the payment plan. It's a hoax. Like fast fashion, consumers are encouraged to "upgrade" to the latest model. But since the launch of the iPhone 1, 2 and 3 – which were hugely different and innovative in performance – technological innovation has slowed, whereas social innovation has flourished. So, don't go for that upgrade after two years; make it last. Ask questions of technology providers. What are their values? And do those values translate into company behaviour? You can vote with your wallet.

You can vote with how you interact with digital technology too, which is wrapped up in the same system of the invisible hand. What is sold as a democratizing movement that connects the world – YouTube, Facebook, TikTok, Twitter – is, in fact, just another form of capitalism where big companies collect your information and monetize it. While I appreciate the advances that these services have made globally, by connecting us and giving people a voice, a lack of regulation and long-term thinking have turned many of these businesses into glorified marketing platforms that trade our data; the last ungoverned domain. If it's free, *you* are the product.

We've all heard how tech executives who have left the industry are famously limiting their children's use of tablets. But the problem runs even deeper than platforms designed to be addictive, which has a significant impact on mental health, not least on children. When technology is designed and programmed, it is done so based on our values. When the values we hold as societies – with our colonial pasts and institutional racism – are subtly translated into our technology, we are excluding huge groups of people. So, we have seat belts designed for a man's body. Automatic soap dispensers that don't take into account dark skin. The message is clear: technology is not neutral. And we need more diversity in tech if we want it to serve society.

When was the last time you entered a building – a shop, a museum, someone's home – and were asked to sign a waiver, to give away all your personal data or divulge where you had just come from and where you were going next? If not never, then certainly rarely. Yet every time we enter an online space, we are being tracked. Not only that, but we often unwittingly give our consent and data to surveillance. Much in the same way as we enter a building, having to accept – or even reject – cookies when we enter a website or app should not be the default option.

On the bright side, there is a growing crop of new technology that is open-sourced and has been designed with a different set of values. Everyone rushed to Zoom when the lockdowns began. We all use WhatsApp. We default to Google. But there are alternatives. Yes, it takes time to find other options, but fortunately there are lots of experts who have done the research. Try the Mozilla Foundation for tech recommendations outside the mainstream that align with your values.

We embrace these tools because they give us so much. But they also take much away. When is enough enough? For me, I would love to completely sideline Facebook, WhatsApp and Instagram but it takes rigour. And the consequence is exclusion. Yet if enough people face that inconvenience, could we turn the tide?

I think so. Organic and vegan options in supermarkets have grown exponentially, from niche to norm. We could see a similar trajectory for technology. So explore the alternative. It might be uncomfortable at first, but good things happen when we leave our comfort zones. Even if the reward isn't immediate, you pave the way for difference, for something new to evolve.

When I turned down the PR job and took Bas's offer instead – Bas, who didn't even own a mobile phone – I did so guided by the belief so accurately expressed by author and activist Alice Walker: "The most common way people give up their power is by thinking they don't have any."

15

Catherine Chong

Investments with Impact

I get asked a lot about "impact investing" by friends. Other synonymous terms include ethical, sustainable and socially responsible, or ESG (ecology, social and governance) investment. People tend to think of philanthropy – donating money to charitable organizations – as impact investing. Not quite. Impact investing is about supporting businesses or social endeavours by providing capital to generate positive ecological and social impacts, with below-market to market-rate returns. Ecological and social gains are seen as part of the returns. Individuals can access impact investing through a number of financial products, including pensions.

In fact, pension funds are one of the most influential ways in which to invest – or divest – your money to good causes.

In Great Britain, private pension wealth was £6.1 trillion in April 2016 to March 2018, up from £3.6 trillion ten years ago.[10] Yet few of us are aware of where our pension providers invest our money. Many are propping up resource-intensive companies, including directly financing the fossil fuels industry; or companies that divert clean water from local communities in water-deprived areas; or businesses that land-grab from vulnerable Indigenous communities, to name a few industry-caused afflictions to ecology and society. To put some of this into context, according to Friends of the Earth, UK local government pensions held £9.7 billion of investments in fossil fuel companies in the 2019/20 financial year.[11] And that's just the fossil fuels industry.

Imagine, instead, if all those billions of pounds, through our pensions, provided much needed capital to address some of the world's most critical challenges: access to clean water, food and nutrition security, innovation in renewable energy, afforestation, the preservation of virgin rainforests. But we don't need to imagine: we can and must make it real.

And it is rather easy to get started. A little online research should get you a shortlist of pension fund providers that say they offer impact investing. Some communicate their investing strategies, the constituents of the fund portfolios, and the quantitative and qualitative ecological and social results of the fund better than others. Ethical exclusion – avoiding the sectors commonly seen as "sin" sectors, such as fossil fuels, tobacco, arms, etc. – is a basic industry practice. More sophisticated funds provide expressed exposure to projects that solve specific ecological or social challenges, to varying degrees. Speak to an impact investing advisor for professional advice. Once you have chosen your pension fund provider, do engage with them on the issues that matter to you. And if your pension is with an employer, show them your research; challenge them to do better; rally your colleagues. Do not be passive.

I believe we should look beyond the traditional, narrow meaning of impact investing to consider how we "invest" our money, full stop. If we think about how our income is spent, most people will find that larger portions are spent on food, clothes and other consumer goods. Who you choose to shop with is a form of impact investing.

My dad was a poultry farmer. Like his parents and grandparents before him, he practised what would be known today as agroecological farming. They learned how to farm in ways that work with nature and animals over many generations. For example, to have a diverse range of crops, trees and animals to support a system of farming that requires no nasty chemicals. When I was young, some men in suits turned up at our farm with the promise of being our best customer and advisor. "We could help you to farm better. Listen to us because we went to university, and you didn't. We can even help you to get loans," was the sales pitch.

It turned out to be a disaster, not only for our family but for our communities. First, they cajoled dad into farming only poultry – the crops, trees and other animals are just a distraction, they said. When they became the only buyer in our town, they forced dad to farm in ways that drastically reduced the wholesale price they paid him – by decreasing the number of rearing days and barn space; using preventive antibiotics to enable the poor rearing conditions; using synthetic fertilizer to reduce the input costs for crops used as feed, and only buying the feed provided by them. And these are just a few of the practices associated with industrial farming.

When all of the 20,000-plus chickens died in the course of a couple of weeks, we were left to deal with the calamity ourselves. Dad believed the feed was the main cause. It was perhaps the last thing that pushed our farm off the cliff of no return. We became very poor. I was lucky to secure a scholarship to study law – a reaction to the injustice we experienced. Later on, I studied climate and socioecological economics to work in the financing of sustainable development. Other than for my passion, the first ten years of my working life were solely about helping to pay back my family's debt.

This is not an isolated anecdote, but one of the many impacts of industrial farming that happened all around the world over the last decades, and continues to happen today. Farmers, their families and the local community have been exploited to create food that degrades nature.

Why do I tell this story? Whenever you shop, you are directing a flow of capital to particular production systems – along with its ecological and social ethos and practices. The food, clothes, furniture, tech gadgets, leisure equipment and services you invest in affect local, regional and global ecology and communities, directly and indirectly.

Many people think that impact investing is for the ultra-rich. That's one of the misconceptions about impact investing. However much you have – whether it's enough to buy three meals a day, to direct your pension fund or to invest regularly in an impact fund – "investing" the cash that you work hard to make is one of the most influential ways you as an individual can effect change, if you care about halting the adverse impacts that trillions of pounds are having on Earth's climate system, ecosystems, biodiversity and people. You have the most incredible power to choose what and who benefits financially – from the roast chicken you eat to the pension you hold.

Three Things to Do

It took me a long time to switch my current account to Triodos Bank. Not because the switch itself was laborious – it was straightforward – but because I wrestled with the decision.

For all the obvious reasons, we place a lot of value on the money we make. I couldn't present a rational argument for why I worried about making the switch, other than the fear of leaving a bank that had kept my money safe for 20 years. As ridiculous as it sounds, I suppose I was worried that Triodos would somehow lose my money. Not because of a poor track record – its reputation is solid – but because of the irrational thoughts we have when it comes to the things we value. Triodos cares about trees more than capitalism!

Fortunately, the embarrassment of not doing so, when also writing a book advising people to switch to an ethical bank, was stronger. I just wish I'd done it sooner.

ACTION 1: BANKING

To do: switch to an ethical bank. It is easy to do and most are protected by the Financial Services Compensation Scheme (FSCS).[12]

- How much do you know about how your bank invests your money? Do some research to find out. *Which?* magazine (UK), Consumer Reports (US) and Choice (Australia) are good places to start.
- Ask your bank directly to detail where your money goes: is it used to fund fossil fuel companies or environmental and social enterprises?
- If the investment funds are unsatisfactory to you, does your bank have plans to change them? Put pressure on high street banks to make the change.

ACTION 2: TECHNOLOGY

To do: explore alternatives in technology. From smartphones and search engines to social media platforms, tech is controlled by the same large companies. But there are alternatives with different ethics; choose those that align with your values.

- Do you need to upgrade your phone? Or does the one you have function well enough?

- If you do have to buy a new phone, research the practices and values of the provider. *Ethical Consumer* magazine offers advice.
- As annoying as it is to read and change your privacy preferences every time you enter a website, it's worth doing so to protect your privacy. At the same time, make informed choices about the digital platforms you choose to engage with. Consult not-for-profit organization Mozilla Foundation.

ACTION 3: IMPACT INVESTMENT

To do: divest your money via impact investment. Options to invest in ecological and social funds via vehicles such as pensions are growing and worth exploring.

- If you have a pension, do you know which funds your money is invested in? Do you know what your options are?
- Ask your provider (whether it's a personal or work pension), consult an advisor and do some research. Do these investments align with your values?
- Use the same approach for all investment opportunities.

BREATHE

- At least half of the Earth's oxygen comes from the ocean – more than from any other source, including trees.[1] Marine plants known as phytoplankton absorb carbon dioxide and release oxygen into the atmosphere. Phytoplankton levels have fallen by 40 per cent since the 1950s.[2]
- Forests cover a third of land across the world and absorb almost one third of the carbon dioxide released from burning fossil fuels every year. Since 1990, it is estimated that 420 million hectares of forests have been destroyed.[3]
- Households consume 29 per cent of global energy and consequently contribute to 21 per cent of resultant CO_2 emissions.[4]

I didn't like sports lessons at school. I once hid behind the ice cream van to miss the sack race on sports day. During cross-country running in secondary school, I found it hard to breathe and often finished last. But swimming was different. I could swim for ages without triggering my childhood asthma. And I liked it.

Ten years ago, I went on a swimming holiday in the Greek Cyclades. Every day, our group – 13 strangers who became friends – would swim an average of 5km, island-hopping from Ano Koufonisi to Schinoussa, to tiny specs of land whose names, if they had one, I can't recall. But I remember our last day so clearly. Treading water, our instructor gathered us together and said, "I want you all to dive now and follow me through an underwater arch." We looked at each other, confused. What arch? We couldn't see any rock formations, just the sea. But we did as we were told – and the arch was duly there. Once through it, we emerged onto a beach. It was spectacular. You wouldn't know it was there – you wouldn't be able to reach it – without swimming through that rocky archway.

What began as a sport that didn't make me breathless in childhood became one of my favourite things to do.

But I didn't realize until recently the extent to which the ocean helps every single one of us to breathe. Trees get all the attention, yet our seas, through photosynthesis, are responsible for at least half of the oxygen in the atmosphere, with some studies suggesting it could be as much as 70 per cent.[5] I found that incredible. Until researching this book, I had no idea of the intrinsic relationship between ocean and oxygen.

I also learned that our oceans' ability to provide us with clean air is being threatened, so I desperately wanted to identify the definitive collective action necessary to fight this risk. But there isn't just one. Not only that, some efforts appear – on the surface – far removed from both oceans and oxygen. As I became aware of the more nuanced actions, I began to appreciate their value. They call for a change in mindset, in behaviour.

Indirect action brings together all three essays in this part of the book. In November 2021, scientists warned that rain could replace snow as the Arctic's most common precipitation. Even if the global temperature rise is kept to 1.5°C or 2°C, the regions of Greenland and the Norwegian Sea will still become rain-dominated. In

August of the same year, rain had fallen on the summit of Greenland's ice cap for the first time on record.[6]

What does this have to do with the air we breathe? Major incidents in the polar regions are the first signs of what is to come for the rest of the world. Dr Emily Shuckburgh's trips to the Arctic and Antarctic have influenced her work on clean energy back in the UK, where home energy use contributes about 15 per cent of all greenhouse gas emissions.[7] The cleaner we can make our energy consumption, the cleaner our air becomes.

After oceans, it is trees that supply us with oxygen to breathe. Known as carbon sinks, forests absorb and store almost one third of the carbon dioxide released from burning fossil fuels every year. But because they also store carbon, deforestation releases greenhouse gases back into the atmosphere. Agriculture is the biggest cause of deforestation, but many industries rely on it for growth. According to Greenpeace, the Amazon rainforest is particularly affected by deforestation. In 2019, almost 10,000km^2 of rainforest was cleared, up by 30 per cent from the previous year – and resulting in the largest loss of the South American rainforest in a decade. This isn't a problem that's far away and far removed: the Amazon is the planet's largest land-based carbon store, which helps in the fight against climate change.[8]

When the problem is so big, it's hard to know what to do. When the consequences are existential, fear often prevails. In situations like this, I turn to scale. Protecting the air we breathe requires a gargantuan effort. We can cut our meat consumption. We can write to powerful supermarkets such as Tesco to demand that they stop using forest-destroying suppliers and replace half the amount of meat they sell with healthy, affordable plant-based alternatives by 2025. We can support organizations such as Greenpeace, who bring these stories to our attention.[9] We can boycott products - such as shampoo, bread, toothpaste and chocolate - that contain palm oil, which comes from the fruit of oil palm trees being cut down to give us these commodities. We can support organizations such as TreeSisters, a reforestation charity that, through individual and business donations, has planted 22 million trees across locations in Brazil, Borneo, Cameroon, India, Kenya, Mozambique, Madagascar, Nepal and West Papua. We can vote for political parties that will use action – not rhetoric – to fight the climate and ecological crises.

Our forests, our oceans, our polar regions play a crucial role in the air we breathe, in shaping the way we live, wherever we are in the world. The actions we take to safeguard them may seem indirect, but they are all connected.

16

Dr Imogen Napper

Turning Ripples into Waves

I've always lived by the sea. I grew up in Clevedon, a small seaside town in the southwest of England, overlooking the Severn Estuary. I spent my childhood exploring the coast and hidden beaches. The mudflats at low tide mean it's no golden stretch of sandy beach, but it's beautiful. And it was always clean – at least that's how I remember it. But as I got older, and started surfing and exploring more beaches in the UK, I began to see plastic waste everywhere. It made me sad. It made me angry. It made me curious.

I'm now a scientist, with a particular interest in marine pollution. We tend to think of trees as the lungs of our planet, but forests are responsible for only about one-third of our oxygen. Most of the oxygen in the atmosphere is produced by marine plants called phytoplankton. I call them the unsung heroes of the world.

As part of my PhD research, I participated in a clothes-washing study: each time you wash your clothes, plastic fibres are released into our rivers and oceans and we wanted to understand what the solutions might be. To continue this research, we received funding from National Geographic, which led to me becoming a National Geographic Explorer and being invited to be part of the expedition team to the Ganges.

We travelled the entire 2,525km of the Ganges, from the Bay of Bengal to the glacial source in the Indian state of Uttarakhand in the western Himalayas. It was beautiful, a river so wide that it looked like the ocean – you couldn't see land from side to side. We met fishermen, discussed our research with the local communities and worked with an international team of inspiring scientists. We wanted to find out how much microplastic – the tiny bits you can't see – was going out to sea, and how much was on the water's surface, in the air, in the sediment. Yes, we were studying the Ganges, but what we found is reflective of any river. The polluted sections, the poor waste management systems – it could have been the Thames. It could have been any river in the world.

When we analysed the results a few months later, we found that, going down the river from the source, the amount of plastic in the water was accumulating and becoming more abundant, and we were able to make some estimations that three billion microplastic particles per day can potentially leave the River Ganges, go into the Bay of Bengal and out into the ocean. That's just one river and associated tributaries.

We made educated guesses as to where these microplastics came from, and we've linked them back to the washing of clothes, to wet wipes and the breakdown of larger plastic items. But there'll be other sources. We believe that more than 90 per cent of the microplastics found were fibres and, among them, rayon (54 per cent) and acrylic (24 per cent) – both of which are commonly used in clothing – were the most abundant.

Plastics such as these have only been manufactured in the last 100 years, yet we've already been able to make this mess. Yes, plastic is an incredibly useful material that has revolutionized our lives in so many beneficial ways. But it says something about our society when our oceans have become plastic soup. We make more than we need, and we can't handle the waste we produce. It is easy to forget about microplastics because we can't see them.

But there was a positive side to our research: we found that small changes can have a huge impact. In 2015, as part of a small research project, we found that there were three million plastic microbeads in one bottle of facial scrub – for every squirt, that's 10,000 particles going down the drain, into our waterways, every time you wash your face. That small piece of research influenced international legislation to ban microbeads in these products. But consumer power also played its part, before the law changed. As people became aware of the findings, they started to make choices: instead of buying their normal brand of facial scrub, they opted for a different one free of microbeads. By doing so, each person prevented three million particles from entering our oceans. Companies listened, voluntarily removing these products from the shelves even before the legislation came in. Consumers have an incredibly powerful voice. So continue to use it; check ingredients, and opt out of plastic packaging if you have the means to do so.

Every time you wash your clothes, up to 700,000 plastic fibres are released in the washing cycle, which could enter our environment. That's a single wash. Now multiply that for your street, a town, a whole city and how many times we wash our clothes per week. By only washing your clothes when you really need to, you could prevent hundreds of thousands of microfibres entering our oceans.

Our research into biodegradable plastics found that they need very specific conditions to break down in the natural environment, so even so-called "better" plastics should ideally be avoided. What's better, in fact, is to reuse whatever plastic bag you currently have when you do your shopping – plastic is an incredible material that will last for years and years.

There is so much we can do, but if I had to pick one piece of advice, it would be this: if you see rubbish on the floor – and it's safe to do so – pick it up and put it in the bin. You will be surprised at how many eyes are on you. It may not be your litter, but it's your planet – our shared planet – and you can take ownership in that moment. And empower others to do the same, to be a good citizen.

You might ask: what possible difference can an individual make to the air we breathe and the environment around us by picking up rubbish? But what if we flipped that question on its head and asked instead: what would happen if everyone picked it up? Our environment would be so much cleaner.

You might also ask: won't that rubbish end up in landfill anyway, regardless of whether or not I pick it up? If we're picking up rubbish off the floor and placing it in the bin, we are taking it out of our environment and putting it into a contained, managed place where waste is destined to go. But beyond that, the more rubbish we pick up, the more we realize how much waste we produce. And, hopefully, the more questions we will ask about how to remedy this.

Being curious led me to a career in scientific research. When we start asking questions, when we discuss our concerns with other people, we turn ripples into waves. We don't need a handful of individuals doing it perfectly. We need an army of people who are trying, willing to learn and doing it imperfectly.

We need to get away from the finger-pointing culture of who is doing it best. Some people will be able to do more depending on their lifestyle, on how much money they earn, on where they live. We're all on different journeys, depending on our means, but with a shared goal. Education is power, and the more education that is given to the consumer, from industry and the government, the more we can achieve – by working together, not separately.

17

Yvonne Aki-Sawyerr, OBE

Breathe the Change

I remember the moment vividly. It was December 2015 and I was driving along Grafton Road, on the outskirts of Sierra Leone's capital city – my hometown – Freetown. I'd driven along that road many times, but I clearly hadn't paid attention to my surroundings. Growing up, I used to see hills covered in lush, green forest. But that afternoon, I didn't see any trees; the once-beautiful hills were barren. I felt sick. I began to grieve, not only for the loss of the natural world, but also for the reality of climate change right before my eyes.

Two years later, a landslide in that area killed more than 1,000 people in less than five minutes. This was not an abstract, isolated tragedy. This is what happens when we lose our forests. Extreme weather patterns – abnormally heavy rainfall or no rain at all – can lead to crop failures and the migration of people from rural areas to cities, putting pressure on infrastructure and leading to deforestation. This was happening in my hometown.

So I decided to do something about it. I ran for mayor of Freetown, a position I've held since 2018. One of my favourite initiatives was to make Freetown a tree town again. One of our goals was to increase vegetation cover by 50 per cent by the end of my term in 2022. That meant we would plant one million trees. For this to work, we needed to involve everyone, to make our city collectively proud of what we can do together. For example, anyone in our town can be a tree steward. But more importantly, every single tree has a unique identifier which can be tracked via a custom-made tree tracker app. We also created 600 jobs for tree growers.

#FreetownTheTreeTown is an example of how a small town like ours can capture the global imagination by doing something that is rooted in community. A million trees will reduce the risk of landslides and flooding in Freetown. They will reintroduce biodiversity. They will protect our water catchments. And it works for us because we are not big emitters of greenhouse gases; we are tiny emitters, unlike the nations in the Global North. But we suffer the most. We need the big emitters to stop burning fossil fuels, to not cut down our planet's trees in the first place. A million trees is our contribution to increasing the much-needed global carbon sink.

But it could only work as part of a structured response. #FreetownTheTreeTown is part of a much wider initiative called Transform Freetown, which aims to address the city's socioeconomic challenges and environmental vulnerabilities. With the principles of strategic planning, systematic implementation and data-driven performance management, we identified 11 priority sectors, grouped into four key clusters – Resilience, Human Development, Healthy City and Urban Mobility – and agreed on 19 measurable targets.

Turning dissatisfaction into action is a summary of my life. My dad tells a story – I must have been about 12 years old at the time – of me seeing a girl selling oranges on the streets in Freetown with a tray on her head, in tat
clothes. I went home, packed my own clothes and

them to her. I had a lovely childhood. My dad is an academic, my mum a teacher. Poverty wasn't something that touched me personally; I was a little bit cocooned from it. We weren't rich, but we were comfortable.

It wasn't until I went to university that I realized not everybody had my life. I met people who had to make choices about whether or not they would eat three times a day. At the same time, I was studying economics and my dissertation was on the impact of devaluation on our economy. I remember reaching out to an International Monetary Fund (IMF) representative for an interview because I was flabbergasted at the introduction of structural adjustment programmes – loans to countries in economic crises. Even at the age of 19, it was obvious to me that this was a failed policy, destined to destroy our economies.

I was born in a poor country that shouldn't be poor. We have diamonds, gold, bauxite, timber, shrimps, lobster, freshwater, saltwater – everything you could possibly imagine. Yet what we have is abject poverty in the midst of wealth. Why? Mismanagement, corruption, collusion with disreputable multinationals. It's fuelling businesses and lifestyles for a small number of people at the expense of the rest of the population.

By all means, plant a tree. Or donate to those doing the work. They're the lungs of the Earth. But go further, too. The biggest difference an individual can make to the climate and ecological crises is to collectively use our power to vote. Planting a million trees in Freetown has taken an enormous, collective effort, underpinned by a structured, political response. So use your vote to support administrations that are committed to making the big changes, to making real interventions, particularly around divesting from fossil fuels.

The steps to address a deep sense of anger and frustration don't magically unfold; that's not how the power of dissatisfaction works. It works when you know things can be done better. And it works when you decide to take the risk to bring about that change.

We all lead different lives; we have different means, different circumstances. But we all have the right to clean air. And we all have one thing in common: we can hold our leaders to account. So make your vote count.

18

Dr Emily Shuckburgh, OBE

Helping the Planet from Home

It began with the robins singing on the rooftops. They'd not been heard before this far north in the Canadian Arctic. Then came the salmon. "We've been fishing them from the Hudson Bay, but they shouldn't be here," a resident of the village of Iqaluit told me, uneasily. This was a decade ago now, and I still remember her cautious summary of events: "It feels like a friend who has suddenly started acting strangely."

The people of Iqaluit are so intimately tied to their environment that they are particularly sensitive to these unsettling changes. And so should we be, because what happens in the polar regions deeply affects the rest of the world. The greatest concern in the Antarctic is for the future of the vast West Antarctic ice sheet – vital infrastructure to our planet. It is not just the ice melting from above, but also the warmer ocean waters getting underneath the ice sheet and melting it from below. This could lead to cliffs of ice simply breaking off. These instabilities could accelerate the disintegration of the ice sheet and the trickle-down effects to the rest of the world.

My trips to the Arctic and the Antarctic have helped inspire me in my role as Director of Cambridge Zero, the ambitious climate change initiative of Cambridge University, to create solutions for a climate-resilient, zero-carbon future. Part of our work focuses on clean energy. With the millions of gas boilers in our homes producing twice as much carbon emissions as all our gas-fired power stations combined, helping to clean up the air we breathe can begin at home.

Much of the action we take is dependent on circumstance. Not everyone owns their home, so there is a limit to what is possible. Some of us are very wasteful when it comes to our energy consumption, but those living in fuel poverty most certainly aren't. There isn't a one-fits-all solution, but in the context of our personal circumstances, most of us can do something. Some actions are simple; others require more effort.

As I write, in mid-November, it is 6°C outside. Apart from the room I'm working in, no other has the heating on, but wearing two jumpers I am perfectly snug. You can turn down your thermostat and replace the valves on your radiators with smart valves to control the temperature in each individual room via an app on your phone. Or you can switch them on and off manually! Similarly, think about your use of hot water: a quick shower uses less water and heat than a bath.

If you have a loft, insulate it, and explore what other options you have for draught-proofing and insulation. This will save energy and money – and create a comfier home. Carefully choose your supplier of energy. And consider switching to technologies such as solar panels, solar water heaters and heat pumps.

Perhaps less well known is the impact of wood-burning at home on both indoor and outdoor air pollution. They seem lovely – cosy, open fires centring a living room. But they have a dark side. They release PM2.5, tiny particles we can't see but which are linked to a range of health conditions, from asthma to an increased risk of lung cancer. In the UK, government research found that more than a third of PM2.5 emissions come from domestic sources, including fireplaces, bonfires and fire pits – a bigger source than vehicles.[10] Even "eco" stoves are problematic. A modern wood-burning stove can emit more PM than an idling diesel lorry.[11] There are healthier ways of heating your home and alternative ways of creating an appealing aesthetic. House plants. Fairy lights, even.

We can make even more of an impact at community level. I live near a village called Swaffham Prior in east Cambridgeshire, whose houses were heated using old oil boilers – dirty, smelly and very polluting. The residents got together to demand a cleaner system – and they are getting it: a heat network with ground and air-source heat pumps that pipes hot liquid directly into homes, where a heat exchanger instead of a boiler heats the water that runs through their existing radiators.

This isn't possible for everyone – Swaffham Prior is a rural village – but what's important is the collective action. Perhaps the neighbours on your street would like to replace their boilers? As a network, you can share information and knowledge. I bet someone is an expert at applying for grants. Another is a brilliant organizer. Someone else knows a trusted contractor. Or perhaps there's a neighbour who has done their homework on renewable energy and will share their findings. Or maybe there's another with a wood-burning stove, unaware of its link to air pollution, with whom you can share your knowledge. Connecting through your community makes the challenge – whatever it might be – easier. We need our global leaders to respond to the climate and ecological emergencies with urgency. But as individuals we can also make a positive difference, especially if we act together.

Every action we take helps to protect our precious polar regions, which in turn, helps to create the carefully balanced environment we need to live our lives. It is all connected. I know the ice sheets covering Greenland and West Antarctica hold within them many metres of sea-level rise that, if released, would eventually devastate coastal communities around the world. I know that only half the autumn Arctic sea ice coverage remains, compared to 40 years ago, when satellite records began. I know the science is irrefutable. But you can abstract yourself from data. It's only when you go there that you appreciate those magnificent natural surroundings, a pristine wilderness and a vastness and scale not seen nor felt anywhere else. And you can't visit these regions without a profound sense that they need protecting. Without a profound sense of responsibility.

We are so used to tomorrow being pretty much the same as today that it's quite difficult to conceptualize what, in the scientific world, we call tipping points – thresholds that, when passed, lead to large-scale and irreversible change. My visits to the Poles have given me some insight into the potential scale of the change.

Perhaps because we can't *see* energy, we've become accustomed to thinking it's just there, that it just happens. We need to get into a mindset of seeing it as a resource, one that needs to be used carefully, rather than as part of a throwaway culture. That mindset, when applied to our daily lives, can make a big difference.

Three Things to Do

When I was about nine years old, my mum took me and my sister to see a friend of hers, who had three daughters. It was one of those awkward moments when you're all expected to play together, despite having never met before.

After a while, it was decided we'd all go for a walk. A few minutes into it, the three girls pulled out plastic bags from their coat pockets and started filling them with rubbish they picked up off the ground. I remember feeling embarrassed. What if one of my friends saw me with children who picked up other people's rubbish? I didn't help them at the time, but the memory has stayed with me ever since. And for years later, when I thought no one was looking, I'd pick up rubbish from the floor and throw it in the bin. I wish I hadn't been so self-aware, but that's a big ask from a child.

Dr Imogen Napper talks about the value of having eyes on you when you pick up someone else's rubbish. She's right. I've never forgotten that day. If I had joined in – publicly at least – perhaps I would have encouraged someone else to follow suit, too.

ACTION 1: PLASTIC WASTE

To do: every time you see a piece of litter on the ground – and it's safe to do so – pick it up and put it in the bin. Not only will you be placing it in a contained environment and setting an example for others to follow, you'll have an even clearer understanding of how much waste we produce.

- When you go shopping, in addition to taking your own bags, do you also take containers or bags for fruit and veg, for the fish counter, or for any refillables?
- Do you wash your clothes more regularly than you need to? Could some items be worn a few more times before they go in the wash?
- How often do you notice litter on the ground? Do you ever pick it up and put it in the bin?

ACTION 2: POLITICAL ACTION

To do: if you care about social and environmental causes, vote for political parties who will act upon them – and don't settle for rhetoric.

- Are you registered to vote?
- Are you registered with charities and organizations that lead national and international campaigns to fight

the climate crisis on our behalf? Signing up to regular newsletters from organizations that fight your causes is a good way to stay informed and lend support.

- Next time you add your name to an open letter or sign up to an organization's newsletter, could you forward it to a friend? Rally people around these causes, too.

ACTION 3: HOME ENERGY

To do: save money and energy by carefully insulating your home and choosing a green energy supplier.

- Do you know how well-insulated your home is? Do a recce. There are many things you can do yourself, including draft-proof windows and doors, stop cold drafts from coming down your chimney; insulate your pipes and water tanks; and even put in your own loft insulation.
- Are you satisfied with your energy supplier? Do some research on green suppliers. Refer to consumer and not-for-profit citizen's advice organizations as a starting point.
- Do you have a wood burner at home? If you're not reliant on it to heat your home, consider getting rid of it or replacing it with a greener energy source.

TRANSFORM

- In a single week in 2019, six million people across time zones and generations took to the streets to demand urgent action against the climate and ecological crises.[1]
- The world's top 100 companies generate more than $15 trillion in revenue each year. As Mark Maslin, Professor of Earth System Science at University College London, says, "Corporations have immense power and we must harness this to change the world for the better."[2]
- Indigenous people protect 80 per cent of the world's biodiversity, which supports all life on Earth.[3] Ecosystems of different species and organisms work together to supply everything we need to survive, from clean water to food and medicine. But Indigenous people remain some of the world's most threatened groups.

"Why are the photographs in black and white? Is it because they were taken in the olden days?" my seven-year-old daughter asked, mesmerized by Sebastião Salgado's portraits of Bela and Manda, with their splendid headdresses and painted faces. She'd neither seen nor heard of Indigenous people before.

"And is that *really* a river?" She questioned the caption next to a photograph of Rio Negro, a major tributary of the Amazon. "You can't see the riverbanks; it looks like the sea."

Amazônia, an exhibition of portraits and landscapes from the Amazon rainforest at the Science Museum in London in 2021, captured my daughter's attention and imagination for an entire afternoon. And her questions summed up my own feelings as I took in the magnificent collection: scale and otherness.

Salgado spent seven years documenting the lives and lands of 12 different Indigenous communities in the Amazon rainforest – the world's largest ecosystem – to tell the story of tipping points: a unique and vital beauty, threatened by the climate crisis.

When Portuguese explorers first discovered Brazil in 1500, the Indigenous population was estimated to number around five million. Today, it numbers no more than 370,000. Ever since its discovery by the Portuguese, this once lush land has been exploited for cattle ranches and soya bean plantations, by logging companies and gold miners. In 2021, a report published by Slow Factory found that more than 100 fashion companies were linked to deforestation of the Amazon through leather supply chains. As land is cleared, so is biodiversity and Indigenous people's homes.[4]

Notes on the *Amazônia* exhibition said that drought and deforestation are threatening the rainforest's ability to provide crucial rain. If rainfall continues to decline, the Amazon's flourishing forest would become a dry savanna, to the point of no return. Its ongoing destruction would release billions of tonnes of carbon dioxide into the atmosphere and lead to climate tipping points that would make parts of the Earth uninhabitable.[5]

I was born and grew up in Brazil but never visited the Amazon. My mum did. In 1978, from Santos, via Bahia and Fortaleza, she reached Manaus, where – she told me – locals set their watches by the weather: "Shall we meet before or after the rain?" they'd say. Recalling her trip along Rio Negro, she added: "You couldn't see further than a metre in depth because the sediments make it so black – it's spectacular. The Indigenous children would wave as we went past."

The 370 million self-identified Indigenous people around the world represent 5,000 different Indigenous cultures and live in at least 70 countries, but make up less than 6 per cent of the global population.[6] Small in numbers, rich in cultural diversity.

But Indigenous lands contain some 80 per cent of the planet's biodiversity, which supports all life on Earth – ours included.

While most humans have been putting increasing pressure on the planet by consuming more resources than are available, Indigenous people have been its caretaker. Through centuries of ancestral knowledge and practices, they have been safeguarding our most fundamental resource. Protect nature, save the planet.

With every interview conducted, every essay written for this book, I've learned something, but Jennifer Martel gave me the history lesson that was never taught. Her story is not an isolated tale. Indigenous people around the world are some of the most vulnerable and marginalized groups, subjected to genocide for centuries.

The three women in this final chapter don't know each other. They haven't read each other's essays. They tackle different subject matters. But when I interviewed Lucy von Sturmer and Gail Bradbrook, they both mentioned – unprompted – the necessary wisdom of Indigenous people around the world.

Perhaps of every solution offered in *Three Things*, the last one matters the most: be a good ancestor to future generations. If you can do that one, you can do them all.

19

Dr Gail Bradbrook

From Grief Comes Courage

If you're feeling panicky and edgy, if you're grieving and fearful, that's an appropriate response. But don't turn away from it, because in those feelings is the salvation of humanity. Grief is the price of love, and from that comes courage.

The climate emergency, the ecological crisis, isn't somewhere else. It's here and now. And it's children who are on the frontline; it's our families in the Global South who are the most impacted and yet did the least to cause it. Based on the science, it's quite possible that civilization will collapse, that 97 per cent of all life on Earth will go, possibly in my children's lifetime. In Indigenous cultures, people want to be good ancestors. They do things for life and think about the next generations.

But in our Western society, we have a systemic problem. Even though we've known about the climate emergency since the 1970s, our linear economy relies on ignoring these facts. Political and economic systems are based on infinite growth on a finite planet; they are hardwired to destroy life on Earth. When 50 per cent of carbon emissions come from 10 per cent of the global population, we need to do more than just cycle to work. Only civil disobedience on a large scale can bring about real change.

Writing to your MP or holding up a placard can add to the noise, but our system is designed to ignore these actions on the whole. When you sit on a road, chain yourself to a building, break a pane of glass, it cannot be ignored. Disruption is part of the mechanism.

But participation in Extinction Rebellion (XR) doesn't require having to break the law. For every person who gets arrested, we need ten others in supporting roles: responding to the media, managing social media, fundraising, organizing wellbeing and support. Some of us are better equipped to deal with it, and people need to think carefully about whether that's right for them. I'm not here to judge. I don't know everyone's situation. You do what you can.

What I would say is that – sometimes, but not always – there's an element of mischief in getting arrested; there's something about the energy in defiance. There is also the spiritual side, a process of initiation. What I mean by that is that there's a part in all of us that is complicit in this crisis. So when we say no to it physically, with our actual bodies, we say yes to life. This is hugely empowering.

Of all the XR movements, blocking the printing presses owned by Rupert Murdoch in September 2020 was beyond joyous. The outrage toward us was incredible. Imagine if we had that level of outrage against all government policies that have taken us in the wrong direction. We were accused of attacking the free press, but our press isn't free. We targeted the billionaire-owned media because they are not responding to the scale and the urgency of the climate and ecological crises, and the main reason for this is that our press is in the hands of the powerful who have vested interests, who are set on dividing us, and are in the pockets of the fossil fuel industry. And we did this – as with every action we take – peacefully, non-violently, with love.

Part of the cycle of civil disobedience is that they come after you; the rebels involved in the blocking of the printing press had their private lives pored over by the media. When the suffragettes were smashing windows, they weren't directly affecting democracy. They were saying: we are angry and we demand our vote. Every other reformist-type approach had been tried – and ignored.

It's unlikely that most of us would become activists through logical argument. We need to wake up, to feel it. Our complicity is our silence. We have to come together and express our power. What we choose to do today is the difference between life and death on Earth. We've got to make something more beautiful.

20

Lucy von Sturmer

Challenge Accepted

When David Attenborough joined Instagram in late 2020 his first caption read: "Saving the planet is now a communications challenge." For him, the decision to join the social media platform – not his "usual habitat" – was likely to be numbers driven, but his appeal should be a call to arms for creative individuals and industries to help drive systemic change.

Language is a powerful tool. It can unite and it can divide. Before launching Creatives for Climates, a peer-to-peer network of global creative activists, I spent a decade working for B Corps[7] such as Dopper and Innate Motion, for NGOs such as Solidaridad and Five Media, and for multi-stakeholder initiatives such as The ZDHC Foundation that brought together civil society and brands to eliminate toxic chemicals from the global fashion supply chain. I had global, high-level exposure to various bodies of "impact" very early on in my career, and an early exposure to the critical role that communication plays.

Discussing campaigns around the table, executives would say, "Phrase it like this, but not with that image", "Would that word alarm donors?" or "Use this word, rather than that word" – referring to nuanced, high-level phrases that had me feeling like an imposter.

The youngest in a room of stuffy decision-makers and their jargon, I would ask myself, "Does anyone really understand what's being communicated?" No, nobody did. More troubling was when I worked briefly in advertising for a global advertising firm, who would employ the same language as NGOs, but without the checks and balances, without the critical thinking, without holding themselves accountable as storytellers, without questioning their position of power.

The nuance of what you say and don't say is highly politicized, with jargon as a weapon, a tactic of delay. For businesses, oftentimes it's intentional to keep things obscure and complicated. But the opposite – language that is clear and clever, with meaning – has the tremendous power to mobilize people to create real change.

Creative for Climates is an offshoot of The Humblebrag, the impact consultancy I founded in 2017 to support brands and business leaders to build their reputation as change-makers. With the Humblebrag, I work with large clients such as the European Parliament, and global agencies such as Danish creative agency &Co, Twenty Twenty, HarrimanSteel and more. But I also wanted to work with impact makers and social entrepreneurs directly on the frontline. To achieve this, at scale, I created a training programme – "Thought Leadership for Change-makers" – to support and amplify the visibility of mission-driven people with new ideas and build a community of frontrunners who, like me, wanted to do more.

Creatives for Climate is a member-driven network I run on the side with a global team, so I can't take responsibility for all the work we've achieved – it's a joint effort from creative individuals from all corners of the world who want to make a difference at a level that drives system changes.

To give you some examples, in 2019, we supported ten different initiatives by making the creative assets for their campaigns – and two of them led to policy change. Fossil Free NL, an Amsterdam-based activist organization, started a citizens' petition to rid the city's advertising spaces of greenwashing campaigns by fossil fuel companies. They asked us to support them with creative executions to make the message hit home. A year later, Amsterdam made history by becoming the first city to ban fossil fuel ads from its public spaces.

Meanwhile, back in New Zealand, the city council of Auckland was about to sign a long-term commitment to use glyphosate and other toxic chemicals as a weed killer in the streets, putting the ecosystem and children's health at risk. We supported the local grassroots organization For the Love of Bees to bring the toxic injustice to light. We called out to the creative community for posters. It was heartening to see the amazing response from creatives around the globe. The creative work hung all over the city, on billboards and in the hands of the activists on the streets. The intervention led to the immediate withdrawal of the plan, keeping the streets of Auckland toxin free.

Perhaps the biggest driver for launching Creatives for Climates was the need to hold myself accountable as an "impact strategist" advising clients on how to build their reputation in this space. What if my dream client turned out to be greenwashing? How can I stay alert to inconsistencies between their actions and words? How can I support them beyond just helping them to gain a competitive advantage in their positioning, but to drive actual meaningful impact in the world?

As the world's best creative minds – in advertising, media, communications – we hold ourselves in high cultural regard; and our values are key to our contribution. However, we can't personally believe one thing and act in another way during our working hours. The time is now to bring your personal – and yes, political – values into the workplace.

More creatives are waking up to their power, platform and influence as storytellers – and choosing to dedicate their talent to driving the change we need to see in the world. The challenge of how to make a shoe brand relevant to teenagers is no longer a cool job. A national airline – no longer a cool brand. At the time of writing, Edelman, one of the world's best-known comms agencies, counts Exxon Mobil as a client – no longer a cool agency.

The change we need is a total system change – and Creatives for Climate is a place to really imagine and act on that. Because everything is connected. The kind of thinking that got us into this mess is not going to get us out of it. We need more feminine values in leadership. We need more people of colour, we need to decolonize our minds, we need more Indigenous wisdom.

It takes work – and an ongoing commitment to be a part of this movement. There are so many levels of privilege, so it's good to get in the habit of asking yourself, "When can I put my voice on hold and pass the mic?" That's where the systems change will start to feel different, because we'll have new voices and new perspectives. And we don't need to help anyone find their voice. They have a voice – we need to give them a platform and move aside.

Above all, my call to arms is for creatives to challenge briefs and to be prepared to say no to work. Agencies thrive off talent, and if talented individuals are no longer prepared to work for companies whose values don't align with their own, that will have an impact further up the chain. Don't be afraid to let people know that you have a policy not to work for businesses linked to the oil and gas industries, for example. And this stance goes beyond the creative sectors; it applies to how we approach work, whatever industry we operate in.

Yes, it demands bravery. Everyone throws around the term "change-maker", but who has the courage to go against the status quo? If it was easy, everyone would do it, but they don't. I remember watching Ben van Beurden, the CEO of Shell, on stage during the Cop26 climate conference in Glasgow in 2021, where he told the story of when his children asked him if he was killing the planet. He told them that the world needs oil and gas right now. Sure, we couldn't transition overnight, but equally, we won't transition at all without disruption. If he really believes he can effect change, let's see the evidence. Is he prepared to live with less? Is he prepared to sacrifice for future generations? Some of us are more to blame than others for the climate crisis, and when the result of the Covid-19 pandemic is an extra 700 billionaires and five million millionaires globally, the system is clearly screwed.[8, 9]

Yet there is a really clear business case for purpose, for being values driven. There's a really clear business case for CEO activism, which trickles down to manager activism, to any level of activism. And it works the other way up, too. A strong business case allows you to have these conversations at work. More scrutiny leads to more action. Good employers will embrace change makers because these are their most passionate individuals. These are the people that are willing to bring their personal values into the workplace.

I didn't invent oil and gas. You didn't either. And neither did Ben van Beurden, but he chooses to go to work at Shell, every day. Those of us in a position to choose where we work hold incredible power. We, as privileged human beings, need to demonstrate more emotional resilience and take responsibility for the choices we make. We need everyone to step up and into this movement.

21

Jennifer Martel

The Bigger Picture

Jennifer Martel is my English name. My Lakota name is Wahukaze Nunpa Win, which means Two Lance Woman. The reason behind it was to be able to live in two worlds from a young age: the Indigenous world and the white man's world. Easier said than done.

In 1868, the US government forcefully relocated the Lakota people to The Cheyenne River Reservation, now home to the federally recognized Cheyenne River Sioux Tribe – where I'm enrolled. I live on the Standing Rock Reservation, on the border of North and South Dakota. But when I was eight years old, I was baptized and taken away from home to live with foster parents in a non-Indigenous family in Idaho. Officially run by the LDS Church from the 1950s until 2000, it was part of the Indian

Placement Programme – another form of genocide, in my opinion – intended to make Native Americans assimilate into mainstream culture via Mormon theology.

I was pretty much the only brown girl in the school and church. I was there for nine years. I had some good times, I had some bad times, I had some really bad times. I missed my parents, my family, my people. I still deal with PTSD and anxiety today. I always said that, when I came of age, I'd leave and come back to help my people. And that's where I'm at now.

In 2018, when we officially gave it a name, I became a founding member of the Indigenous Peoples Movement, a collective of Indigenous peoples from all over the world, united against issues that directly affect our lands, peoples and respective cultures.

Our first march was in January 2019. We weren't marching for a specific issue; we were marching for everything. For awareness. For climate change. For missing and murdered Indigenous women. For the rate of Indigenous men going to prison. For genocide. The Doctrine of Discovery has been used for centuries to legitimize the colonization of

Indigenous land and people. We're the forgotten race. But we are still here. We have a voice and we will use it.

The Indigenous Peoples Movement was born out of the Standing Rock Movement in 2016. We'd heard that an underground oil pipeline – the Dakota Access Pipeline – was to be constructed and we were concerned about its impact on the environment, particularly on our region's water, and the threat to our sacred and ancient burial grounds.

So our young people started running. First, they ran within the reservation. Then, they ran across the river. Our youth ran for 2,000 miles, from North Dakota to Washington DC. With the support of the #NoDAPL social media campaign, the youth ran to deliver a petition with 160,000 signatures, including Leonardo di Caprio's, to the US Army Corps of Engineers against the construction of the pipeline.

Meanwhile, we built a camp at Standing Rock, which grew to thousands of people, all united in demonstrating their opposition to the pipeline. Some said they'd come for the day, and stayed for a week. Some even quit their jobs. They came from everywhere – from Minneapolis, from Seattle, from different countries. We had monetary donations, we had letters of support. We built a second-hand store. We built an information centre. We created eight kitchens. We cooked. We innovated. We bartered and traded. You could come to camp and learn about other tribes' languages and culture. So many tribes, in one area. Everybody getting along, feeding each other, looking after each other, sharing.

Just like we used to, a long time ago. The energy at the camp was ... one day, I was standing there and I just cried. Is this how it used to be? Why can't it be like this – again, always? We have the knowledge, we have the traditions, we have the language; it's still within us. We're still here. It was awesome, it was overwhelming, it was good.

Around the world, some people got it and some people didn't. The pipeline got through. We just don't have the privileges of a white person. We didn't have enough. We didn't have enough to stop it.

We come from a culture of oral history, where a generation is defined as 70 years. In our generation, we were told of a prophecy: the time will come when non-natives will go to the Indigenous people to ask us, "How do we survive this world? How do we save the world?"

If you don't know the answer, visit an Indigenous community. Sit with us; you'll learn a lot. We talk about the good, the bad and the ugly. We talk about life on the reservation and about our interactions with non-Indigenous societies. Some people, once they come to visit us, say, "I want to go back and tell the truth." Tell the truth. It's all we ever wanted. But we're being lied about throughout history. We may not live in teepees anymore, but we do have teepees. We may speak two languages, but we still have our language. We still have our culture. And we're willing to share it. We're storytellers.

Until you come to visit us, in the meantime, here's my answer to how you can save the world. We call it "wotakuye waste" – it means "being a good relative" – and it is our greatest responsibility. To be compassionate, to share, to respect, to care for all living things – the winged, two-legged, four-legged, the river and the fish, the plants, the earth. The Earth. We see the big picture. As Indigenous people, we want to live a good life because we think of the generations coming behind us. Do you?

Three Things to Do

My local park is home to hundreds of red and fallow deer. I seek them out during lunchtime runs or cycle rides. Every time, without fail, they make me a bit emotional. I find them so beautiful and otherworldly.

That has become my default reaction to many things in recent years: birdsong at dusk; blossom in spring; colourful carpets of autumn leaves. I wonder if I feel a need to observe these things, to make them have a physical effect on me for fear that they may no longer exist at some point in time.

Knowing what I know about the climate and ecological crises, it's hard not to feel what I can only describe as grief at times. But I have found a solution: action.

Doing nothing makes me feel powerless and defeated; doing something gives me focus and hope. It shows me how much is possible. Writing *Three Things* is my way of helping to make a difference.

ACTION 1: ACTIVISM

To do: become an activist. Participation in activism doesn't require having to break the law; there are plenty of ways to support activist organizations. Do what you can.

- What does activism mean to you?
- Within a spectrum that encompasses protesting, responding to the media, managing social media, fundraising etc., could you support activism?
- Which activist organizations best reflect your values? Could you sign up to them to determine how you could help?

ACTION 2: ON THE JOB

To do: if you're in a position to choose, say no to work that doesn't align with your values and speak up in your workplace to encourage change.

- What are you good at? What do you love? Could you lend your skills to mission-driven organizations?
- Could you spare one day a week, a month or a year to support such organizations?
- Does your job reflect your personal values and morals? If not, could you make changes to your work or seek advice to help you do so?

ACTION 3: GOOD ANCESTORS

To do: be a good ancestor. Leave the world in a better place than you found it.

- Do you surround yourself with people who share your views? People who live in similar circumstances? How big is your circle? Could it be widened?
- Do you talk about the climate crisis with others?
- How would you like your generation to be remembered?

Epilogue

Every week during the writing of this book, I'd face the same fear: was I delivering on my promise to the reader? I had every confidence in the contributors, but I worried about the premise of 21 "simple" ideas. How could solutions to the climate crisis ever be simple?

As a final check, I made a new list – one based on the efforts of this book's 21 contributors:

- Fight for inquest into daughter's death.
- Prevent landslides by rallying a town to plant one million trees.
- Lead a march against the genocide of your own people.
- Get arrested to protect the future of generations.

When I compiled a list of some of the experts' endeavours, the 21 actions given in the appendix of *Three Things to Help Heal the Planet* didn't look so demanding after all.

Our ability to act collectively on these solutions will depend on circumstance, but there comes a point when we must distinguish between our means and our will. If we're lucky enough to choose how we live, then how many of our actions – or inaction – are determined by our reluctance to give up our comforts?

At the time of writing, British MPs voted to reinstate some parts of the Police, Crime, Sentencing and Courts Bill, after previously suffering a defeat in the House of Lords over the plans to clamp down on people's right to protest. The government's proposal was largely in response to the actions in recent years of environmental protesters who have blocked roads and glued themselves to buildings. What caught my attention above all else was the disproportionate rage that ensued, with puerile language, including "uncooperative crusties" and "heaving hemp-smelling bivouacs",[1] being used to support a political argument. The protesters, who harmed no one in their activism, were putting themselves at risk of being arrested to fight for a collective cause: our future on Earth. Yet we couldn't put up with a delay to our car journeys or with being denied temporary access to a building.

I haven't ticked off all 21 solutions offered in this book yet. That was another fear: the hypocrisy of offering answers without having first completed them myself. But what was the alternative? Some solutions I discovered while writing *Three Things*. Toward the end of the editing process, I found more, which I haven't managed to include. If I waited to complete them all first, another set would probably come along. And then what? Another excuse to do nothing? To stay in my comfort zone?

In writing *Three Things to Help Heal the Planet*, I wanted to solve the problem of inaction with tangible solutions. I wanted simplification, not necessarily of the actions themselves, but in the reduction of distraction from opinion, debate, talk. I wanted to cut through the noise to find out if, on the advice of experts, I could contribute positively to a complex crisis. The 21 contributors in this book have done the hard work, they've put in the hours. What they advocate, in the big scheme of things, is simple. The next step is up to us.

The 21 Solutions

Here's a round-up of all the suggested solutions in this book.

LIVE

- Buy fewer, better everything – fewer goods, fewer services, fewer consumer experiences.
- Conduct a bathroom audit. Place all personal care products – from make-up to moisturizers – into category piles to determine what you do and don't use. Once it's time to replace a product, do so with a brand that cares about its impact on people and the environment.
- Grow some of your own food. You won't be self-sufficient, but you'll learn and appreciate the skill and complexity of growing food – and be responsible for that harvest.

EAT

- Don't throw food away if it can be eaten. Freeze it or share it – apps such as OLIO connect local people to share uneaten food. And buy less for your own consumption next time.

- Create a weekly meal plan. By planning your meals a week ahead, you can shop and cook more efficiently, saving money and reducing food waste.
- Reduce your meat consumption. If you can, go as far as Jonathan Safran Foer's suggestion: don't eat animal products before dinner.

WEAR

- Satisfy the thrill of new clothes by borrowing, renting or buying second-hand instead.
- Start shopping from your wardrobe first. You never know, you might find forgotten favourites or reinvent new looks.
- If you do buy new clothes, do your research and support brands that care about the people and resources behind the making of your clothes.

TRAVEL

- If you are physically able, for journeys under five miles walk, cycle or take public transport.
- Fly less and travel better. Do the research into destinations, airlines and accommodation providers before booking your next holiday.

- If you can, choose alternatives to private cars that run on petrol and diesel.

SPEND

- Switch to an ethical bank. It is easy to do and most are protected by the Financial Services Compensation Scheme (FSCS).
- Explore alternatives in technology. From smartphones and search engines to social media platforms, tech is controlled by the same large companies. But there are alternatives with different ethics; choose those that align with yours.
- Divest your money via impact investment. Options to invest in ecological and social funds via vehicles such as pensions are growing and worth exploring.

BREATHE

- Every time you see a piece of litter on the ground – and it's safe to do so – pick it up and put it in the bin. Not only will you be placing it in a contained environment and setting an example for others to follow, you'll have an even clearer understanding of how much waste we produce.

- If you care about social and environmental causes, vote for political parties who will act upon them – and don't settle for rhetoric.
- Save money and energy by carefully insulating your home and choosing a green energy supplier.

TRANSFORM

- Become an activist. Participation in activism doesn't require having to break the law; there are plenty of ways to support activist organizations.
- If you're in a position to choose, say no to work that doesn't align with your values and speak up in your workplace to encourage change.
- Be a good ancestor. Leave the world in a better place than you found it.

Acknowledgements

You wouldn't be reading this book if it wasn't for the generosity and wisdom of 21 people who gave me their time, answered my questions with endless patience and taught me something new every day. People warned me that writing a book that depended heavily on the contribution of others would be challenging – the 21 experts in this book made it a pleasure.

Thank you Alasdair, Catherine, Chris, Claire, Emily, Eshita, Gail, Imogen, James, Jennifer, Jona, Jonathan, Juliet, Khandiz, Lucy, Melissa, Rosamund, Tamsin, Tessa, Tristram and Yvonne.

Thank you to Charlie for believing in it and to Jo for making it happen.

Thank you to my friends Keely and Laura for that night on the Southbank – I hadn't finished one idea, but they put me onto the second. The belief and efficiency.

Thank you to JP Watson for seeing it first.

Thank you to Roly for the idea. And for encouraging me to steal it.

References

INTRODUCTION: FIRST THINGS FIRST

1. Christiana Figueres and Tom Rivett-Carnac, *The Future We Choose, Surviving the Climate Crisis* (Manilla Press, 2020), p.5.
2. Lily Cole, *Who Cares Wins: Reasons for optimism in our changing world* (Penguin Life, 2020), p.4.
3. Thomas Crowther, Nature-based solutions in the fight against climate change | Thomas Crowther | TEDxLausanne, TED, September 2019. Available at: www.ted.com/talks/thomas_crowther_nature_based_solutions_in_the_fight_against_climate_change?language=en
4. Simon Murphy, "More than 500,000 people sign up to be NHS volunteers", *Guardian*, 25 March 2020. Available at: www.theguardian.com/world/2020/mar/25/astonishing-170000-people-sign-up-to-be-nhs-volunteers-in-15-hours-coronavirus

5. Linda Geddes, "Britons cut meat-eating by 17%, but must double that to hit target", *Guardian*, 8 Oct 2021. Available at: www.theguardian.com/food/2021/oct/08/cuts-uk-meat-consumption-doubled-health-researchers-food
6. Figueres and Rivett-Carnac, p.8.
7. Cole, p.7.

LIVE

1. Stefan Giljum, Friedrich Hinterberger, Martin Bruckner, et al., *Overconsumption? Our use of the world's natural resources* (Friends of the Earth, 2009), p.3. Available at: friendsoftheearth.uk/sites/default/files/downloads/overconsumption.pdf
2. Olivia Young, "Beauty Brands are Going Anhydrous to Combat the Global Water Crisis", Eco-Age website, 22 March 2021. Available at: eco-age.com/resources/beauty-brands-are-going-anhydrous-to-combat-the-global-water-crisis
3. Martien van Nieuwkoop, "Do the costs of the global food system outweigh its monetary value?", World Bank Blogs, 17 June 2019. Available at: blogs.worldbank.org/voices/do-costs-global-food-system-outweigh-its-monetary-value
4. See the Friends of the Earth website: friendsoftheearth.uk/consumption-natural-resources

EAT

1. United Nations, "Food systems account for over one-third of global greenhouse gas emissions". Available at: news.un.org/en/story/2021/03/1086822
2. Feeding the 5000, "Public feasts to showcase the delicious solutions to food waste", Feedback website, 24 August 2017. Available at: feedbackglobal.org/campaigns/feeding-the-5000/
3. Kelly Oakes, "How cutting your food waste can help the climate", BBC.com, 26 February 2020. Available at: www.bbc.com/future/article/20200224-how-cutting-your-food-waste-can-help-the-climate.

4. While the dominant man-made greenhouse gas is carbon dioxide (CO_2), which is emitted when we burn fossil fuels, other greenhouse gases also contribute to climate change, including methane and nitrous oxide. So instead of detailing the breakdown of a carbon footprint, an equivalent – known as CO_2e – is often used instead.

5. For more information, see "Wasting Food Feeds Climate Change", WRAP website. Available at: wrap.org.uk/media-centre/press-releases/wasting-food-feeds-climate-change-food-waste-action-week-launches-help

6. See: The Food and Agricultural Organization of the United Nations, "Key Facts and Findings", FAO website. Available at: www.fao.org/news/story/en/item/197623/icode
7. Paul Mundy (ed.), *Meat Atlas*, Heinrich Böll Shiftung Foundation and Friends of the Earth (2021). Available at: friendsoftheearth.eu/wp-content/uploads/2021/09/MeatAtlas2021_final_web.pdf
8. Jonathan Safran Foer, *We Are The Weather: Saving the Planet Begins at Breakfast* (Hamish Hamilton, 2019), p.51.

WEAR

1. Nathalie Remy, Eveline Speelman and Steven Swartz, "Style that's sustainable: A new fast-fashion formula", McKinsey Sustainability website, 20 October 2016. Available at: www.mckinsey.com/business-functions/sustainability/our-insights/style-thats-sustainable-a-new-fast-fashion-formula
2. Abigail Beall, "Why clothes are so hard to recycle", BBC website, 13 July 2020. Available at: www.bbc.com/future/article/20200710-why-clothes-are-so-hard-to-recycle
3. The United Nations Economic Commission for Europe, "Fashion is an environmental and social emergency, but can also drive progress towards the Sustainable Development Goals", UNECE website, 1 March 2018. Available at: unece.org/forestry/news/fashion-environmental-and-social-emergency-can-also-drive-progress-towards
4. Environmental Audit Committee, "Fixing fashion: clothing consumption and sustainability", Gov.uk, 19 Feb 2019. Available at: publications.parliament.uk/pa/cm201719/cmselect/cmenvaud/1952/report-summary.html

5. United Nations, "Fashion Industry, UN Pursue Climate Action for Sustainable Development United Nations", United Nations website, 22 Jan 2018. Available at: unfccc.int/news/fashion-industry-un-pursue-climate-action-for-sustainable-development
6. F. De Falco, E.Di Pace, M. Cocca, et al, "The contribution of washing processes of synthetic clothes to microplastic pollution", *Sci Rep* 9, 6633 (2019). Available at: doi.org/10.1038/s41598-019-43023-x
7. thredUP, "2021 Resale Report", thredUP website. Available at: www.thredup.com/resale/#size-and-impact
8. Ellen MacArthur Foundation, "A New Textiles Economy: Redesigning fashion's future" (2017). Available at: www.ellenmacarthurfoundation.org/publications

TRAVEL

1. The World Health Organization, "Air pollution", WHO website. Available at: www.who.int/health-topics/air-pollution#tab=tab_1
2. The Aviation Environment Federation, "What Can I Do as a Leisure Traveller?", AEF website. Available at: www.aef.org.uk/action/as-a-traveller
3. The World Health Organization, Europe, "What are the effects on health of transport-related air pollution?", WHO website. Available at: www.euro.who.int/en/data-and-evidence/evidence-informed-policy-making/publications/hen-summaries-of-network-members-reports/what-are-the-effects-on-health-of-transport-related-air-pollution
4. Quoted in Chunka Mui, "3 Reasons There Might Be No Path to Success on Climate Change", Forbes website, 1 Oct 2019. Available at: www.forbes.com/sites/chunkamui/2019/10/01/3-reasons-there-might-be-no-path-to-success-on-climate-change
5. Mums For Lungs, "What is Air Pollution?", Mums for Lungs website. Available at: www.mumsforlungs.org/about-air-pollution
6. The Aviation Environment Federation. Available at: www.aef.org.uk/2019/05/02/ccc-net-zero-report-

well-still-be-flying-in-2050-but-government-can-no-longer-ignore-aviation-emissions-in-its-climate-policies/7. The Aviation Environment Federation (AEF). Available at: www.aef.org.uk/what-we-do/climate

7. King's College, London, "Mixed pollution results for London during lockdown", King's College, London website, 6 May 2020. Available at: www.kcl.ac.uk/news/mixed-pollution-results-london-during-lockdown
8. "Walking and cycling in England", www.parliament.uk. Available at: publications.parliament.uk/pa/cm201719/cmselect/cmtrans/1487/
9. Lord Deben, "Independent advice to government on building a low-carbon economy and preparing for climate change", The Committee on Climate Change, 24 Sept 2019. Available at: www.theccc.org.uk/wp-content/uploads/2019/09/Letter-from-Lord-Deben-to-Grant-Shapps-IAS.pdf
10. The Aviation Environment Federation (AEF). publications.parliament.uk/pa/cm201719/cmselect/cmtrans/1487/148705.htm
11. World Travel & Tourism Council, "Latest research from WTTC shows a 50% increase in jobs at risk in Travel

& Tourism", World Travel & Tourism Council website, 25 March 2020. Available at: wttc.org/News-Article/Latest-research-from-WTTC-shows-a-50-percentage-increase-in-jobs-at-risk-in-Travel-and-Tourism

12. J. Nawijn, M.A Marchand., R. Veenhoven, et al., "Vacationers Happier, but Most not Happier After a Holiday", *Applied Research Quality Life* (2010)5, 35–47. Available at: doi.org/10.1007/s11482-009-9091-9
13. *The Economist*, "The perilous politics of parking", The Economist website, 8 April 2017. Available at: www.economist.com/leaders/2017/04/06/the-perilous-politics-of-parking

SPEND

1. Banking on Climate Chaos, "Banking on Climate Chaos: Fossil fuel finance report 2021", Banking on Climate Chaos website, 21 March 2021. Available at: www.bankingonclimatechaos.org
2. Fairphone, "Longevity", Fairphone website. Available at: www.fairphone.com/en/impact/long-lasting-design
3. United Nations University, "Global E-Waste Surging: Up 21% in 5 Years", UNU website. Available at: unu.edu/media-relations/releases/global-e-waste-surging-up-21-in-5-years.html
4. The Right Hon Alok Sharma MP and Cabinet Office, "The need for pension funds to invest in a clean, green and prosperous future UK Government", Gov.uk, 1 June 2021. Available at: www.gov.uk/government/speeches/the-need-for-pension-funds-to-invest-in-a-clean-green-and-prosperous-future
5. Ric Lander, Friends of the Earth Scotland et al, "Polluted Pensions?: Clearing the air around UK pensions and fossil fuels" (UK Divest, 2021). Available at: www.divest.org.uk/wp-content/uploads/2021/10/Polluted-Pensions-Final-1.pdf

6. Big Livestock Team, "Butchering the Planet: The big-name financiers bankrolling livestock corporations and climate change" (Feedback, 2020). Available at: feedbackglobal.org/butchering-the-planet
7. Banking on Climate Chaos, op. cit.
8. Milton Friedman, "The Power of the Market – The Pencil". Clip available on YouTube.com.
9. Office for National Statistics, "Pensions, savings and investments", Office for National Statistics website. Available at: www.ons.gov.uk/peoplepopulationandcommunity/personalandhouseholdfinances/pensionssavingsandinvestments
10. Friends of the Earth. Available at: friendsoftheearth.uk/climate/dirty-pensions-ps10-billion-still-invested-fossil-fuels-local-government-pension-funds
11. *Ethical Consumer Magazine*, "Three ethical issues to think about when switching your bank account", Ethical Consumer website, 28 July 2021. Available at: www.ethicalconsumer.org/money-finance/three-ethical-issues-think-about-when-switching-your-bank-account

BREATHE

1. NOAA, "How much oxygen comes from the ocean?", National Ocean Service website. Available at: oceanservice.noaa.gov/facts/ocean-oxygen.html
2. Lauren Morello, "Phytoplankton Population Drops 40 Percent Since 1950", Scientific American website, 29 July 2010. Available at: www.scientificamerican.com/article/phytoplankton-population
3. Greenpeace, "Forests", Greenpeace website. Available at: www.greenpeace.org.uk/challenges/forests
4. United Nations, "Goal 12: Ensure sustainable consumption and production patterns", United Nations website. Available at: www.un.org/sustainabledevelopment/sustainable-consumption-production/
5. *National Geographic*, "Save the Plankton, Breathe Freely", National Geographic website. Available at: www.nationalgeographic.org/activity/save-the-plankton-breathe-freely
6. Damian Carrington, "Rain to replace snow in the Arctic as climate heats, study finds", *Guardian*, 30 Nov 2021. Available at: www.theguardian.com/environment/2021/nov/30/rain-replace-snow-arctic-climate-heats-study

7. Damian Carrington, "UK's home gas boilers emit twice as much CO_2 as all power stations – study", *Guardian*, 29 Sept 2021. Available at: www.theguardian.com/environment/2021/sep/29/uks-home-gas-boilers-emit-twice-as-much-co2-as-all-power-stations-study
8. Greenpeace, "Deforestation", Greenpeace website. Available at: www.greenpeace.org.uk/challenges/forests/deforestation/
9. Chiara Vitali, "Tesco's deforestation claims are misleading the public", Greenpeace website, 14 August 2020. Available at: www.greenpeace.org.uk/news/tesco-deforestation-meat-adverts/
10. National Statistics, "Emissions of air pollutants in the UK – Particulate matter (PM10 and PM2.5)", UK Government website. Updated 26 February 2021. Available at: www.gov.uk/government/statistics/emissions-of-air-pollutants/emissions-of-air-pollutants-in-the-uk-particulate-matter-pm10-and-pm25
11. Julia Horton, "How cities draw the heat", Natural Environment Research Council website, 22 Feb 2018. Available at: nerc.ukri.org/planetearth/stories/1885/

TRANSFORM

1. Matthew Taylor, Jonathan Watts and John Bartlett, "Climate crisis: 6 million people join latest wave of global protests", *Guardian*, 27 Sept 2019. Available at: www.theguardian.com/environment/2019/sep/27/climate-crisis-6-million-people-join-latest-wave-of-worldwide-protests
2. Professor Mark Maslin, *How to Save Our Planet: The Facts* (Penguin Life, 2021), p.120.
3. Kanyinke Sena, "Recognizing Indigenous Peoples' land interests is critical for people and nature", WWF website, 22 Oct 2020. Available at: www.worldwildlife.org/stories/recognizing-indigenous-peoples-land-interests-is-critical-for-people-and-nature#:~:text=By%20fighting%20for%20their%20lands,they%20have%20lived%20for%20centuries
4. Stand.earth Research Group, "Nowhere to Hide: How the Fashion Industry is linked to Amazon Rainforest Destruction", Slow Factory website. Report available at: slowfactory.earth/supplychange.
5. See *Amazônia* exhibition details and blogs posts available at: www.sciencemuseum.org.uk/see-and-do/amazonia

6. Department of Economic and Social Affairs, "State of the World's Indigenous Peoples" (United Nations, 2009). Available at: www.un.org/esa/socdev/unpfii/documents/SOWIP/en/SOWIP_web.pdf
7. A for-profit company certified by B Lab, a non-profit organisation that measures social and environmental performance against the standards in the online B Impact Assessment.
8. R. Sharma, "The billionaire boom: how the super-rich soaked up Covid cash", *Financial Times*, 14 May 2021. Available at: www.ft.com/content/747a76dd-f018-4d0d-a9f3-4069bf2f5a93
9. BBC, "Millions become millionaires during Covid pandemic", BBC website, 23 June 2021. Available at: www.bbc.co.uk/news/business-57575077

EPILOGUE

1. Sarah Marsh, "Extinction Rebellion activists glue themselves to DfT and Home Office", *Guardian*, 8 Oct 2019. Available at: www.theguardian.com/environment/2019/oct/08/extinction-rebellion-activists-glue-themselves-to-home-office-and-dft

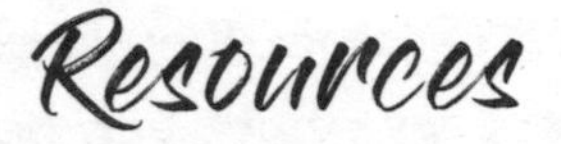

Here is a select list of publications, websites and podcasts to help inspire you.

BOOKS

Tamsin Blanchard, *Green is the New Black: How to Save the World in Style* (Hodder, 2008).

Lily Cole, *Who Cares Wins: Reasons for Optimism in our Changing World* (Penguin Life, 2020).

Christiana Figueres and Tom Rivett-Carnac, *The Future We Choose, Surviving the Climate Crisis* (Manilla Press, 2020).

Melissa Hemsley, *Eat Green: Delicious flexitarian recipes for planet-friendly eating* (Ebury, 2020).

Juliet Kinsman, *The Green Edit: Travel* (Ebury Press, 2020).

Yuval Noah Harari, *21 Lessons for the 21st Century* (Vintage, 2019).

JB MacKinnon, *The Day the World Stops Shopping* (The Bodley Head, 2021).

Professor Mark Maslin, *How to Save Our Planet: The Facts* (Penguin Life, 2021).
Claire Ratinon, *How To Grow Your Dinner Without Leaving the House* (Laurence King Publishing, 2020) and *Unearthed: On race and roots, and how the soil taught me I belong* (Chatto & Windus, 2022).
Jonathan Safran Foer, *We Are The Weather: Saving the Planet Begins at Breakfast* (Hamish Hamilton, 2019).
Alisa Smith and JB MacKinnon, *The 100-Mile Diet: A Year of Local Eating* (Vintage, 2007).
Tristram Stuart, *Waste: Uncovering the Global Food Scandal* (Penguin, 2009).

WEBSITES

Client Earth: www.clientearth.org
Farms to Feed Us: farmstofeedus.org
Friends of the Earth: foe.org
Greenpeace: www.greenpeace.org.uk
TreeSisters: treesisters.org

PODCASTS

What If We Get It Right? with Tessa Wernink
Who Cares Wins with Lily Cole

NOTES

NOTES

NOTES

NOTES

NOTES

NOTES

NOTES